Earth-Keeper Chronicles:
METATRON SPEAKS

To SANDRA.
WITH LOVE,
TYBERONN

by James Tyberonn

THE NEXT 2-YEARS WILL OPEN A
GREAT potential for you to MOVE
forward. It is A great ADVANCEMENT
A quantum leap. But you must
pursue it, create the RESONANCE
by intent AND ACTION. You ARE A
Great SOUL, AND completion AWAITS you.

Earth-Keeper Chronicles: METATRON SPEAKS
by James Tyberonn
ISBN: 978-0-9767035-6-3
Library of Congress Control Number: 2008943153

To order books:
Email: Tyberonn@Earth-Keeper.com,
Website: http://www.Earth-Keeper.com

Altered States, Biographies & Personal Experience, Body Mind Spirit, Energy Science, Gaia, Source Alignment, Ascension, Mastery, Light, Chakras, Channeling, Consciousness: Awareness & Expansion, Consciousness Physics, Crystals, Life Energy, Grids, Sacred Geometry, Sacred Sites, Portals, Everyday Spirituality, God, New Age, Science & Religion, Spirituality, The Self.

STAR QUEST PUBLISHING
RENO, NV PHOENIX, AZ
New Perspectives in Unified Consciousness.

3030 E. Shangri-La Rd., Phoenix AZ 85028
info@StarQuestPublishing.com 602-482-1568
www.StarQuestPublishing.com/index.htm
Printed in China.

Dedication to Ronna Herman

This book is dedicated in love and gratitude to my mentor and spiritual mother, Ronna Herman, who embodies the Divine Feminine in strength, nurturing, unconditional love and wisdom.

Thank you Ronna.

Foreword by Ronna Herman

Even though James Tyberonn has only been channeling Metatron since mid 2007, his messages and rapidly growing following are quite astounding. I am very happy to see that the Earth-Keeper.com website is averaging over one million hits per month. One week alone in March revealed 1.2 million log ons, which proves that the vital information he is bringing forth from Archangel Metatron is of vast importance and is greatly sought after. This is only one of the signs of the times. People are thirsty for spiritual guidance and information. They are growing more desperate every day as the world economy crumbles and as our old reality gradually dissolves in order to be replaced with a new paradigm of unity consciousness, self mastery and personal responsibility.

But James Tyberonn's success is not an overnight feat. His life has been one of intent, focus and disciplined spiritual training that has led him to this shining phase. He worked hard to get here. He grew up in the crystal fields of Arkansas and began traveling the world as a geologist and engineer in the 1970s. He has lived and worked abroad for 30 years. He speaks four languages and has developed a keen knowledge of sacred sites, visiting every continent on the planet and seeking to understand the complex mechanics of the Earth's ley lines, vortexes and portals from both a metaphysical and a scientific stance. During his nine years of living in South America, he read every metaphysical book he could get his hands on and studied gemology and crystallography during his six years in Brazil. It is not lost on me that he grew up in Arkansas amidst the largest quartz formations in the northern hemisphere, and he then spent the next six years in Brazil amidst the largest quartz formations in the world. The divine planning in that is quite apparent, because Tyb is one of the world's best authorities on healing gems and quartz crystals.

His advanced understanding of crystallography has led him to an expertise in what Metatron terms as Phi Crystals in his work with Vogel Crystals, truly Atlantean technology.

Tyb is truly a 'Citizen of the World' having lived in India, Brazil, Africa, Arabia, Europe and the Far East as well as having visited and worked in over 55 countries across the globe. He moved back to the USA in the mid 1990s and began exploring his country of birth. During that process he became involved in Native American spirituality, and he participated in six five-day prayer fasts called visions quests that took him to the next level. His five-day fasts taught him how to communicate directly with the angelic realm and how to achieve out of body travel. These feats were not easy, but he accomplished them. Soon afterward he began working with Native American Elders and South American Master Shaman, and he experienced some transformational shamanic journeys which he described as geometric and multi-dimensional.

Although Tyb is in a large physical frame, at 6 feet 5 inches and 250 pounds, he is one of the most gentle and giving persons I have ever met. He is often called the 'Gentle Giant,' and I am aware that he requires that large energy in order to carry the resonance of Metatron. Within his field is encompassed the energy of every sacred site, vortex and portal he has visited across the planet. He was one of the Golden Tall Beings in Atlantis during his incarnations as Tyberonn, the Atlantean Crystal Grid Master. I met Tyb in 2005, and we recognized one another immediately. Michael told me that I was meeting soul family, and it has been a pleasure to watch him grow into the energy, the enormous energy and sacred spiritual technology of Archangel Metatron. The energy of Archangel Metatron is very intense, very powerful, and only a few can truly hold this energy within them. It has taken Tyb a lifetime, but he now holds and carries the energy of

Archangel Metatron with humility, beauty and spiritual balance. Metatron's messages that come through him are amazing and are a gift to the world. Even though the messages are very scientific and technical, Tyb's ability to interpret them in layman terms makes this vast knowledge of our Cosmology available to everyone from the layman to the advanced academic.

As the Creator Light impacts more and more onto the shadow lands of the Earth and into the hearts of humanity, the desire for true self empowering guidance and solutions which can benefit everyone will be sought after. The old fear based negative vibrational patterns of the mass consciousness will then slowly dissolve.

Most people have no idea what is involved in becoming a Cosmic Telepath, or a person who agrees to bring forth information / teachings from the higher realms of existence. First, before incarnating in this lifetime, we made an agreement to become a cosmic correspondent of solar system, galactic, universal and omniversal information. This information may contain universal laws, our origins, our many journeys throughout this universe and even beyond. It may also contain sacred geometry, the history of the Earth, the sacred sites of past, present and future and the vital role we have to play in this exciting journey into a whole new reality.

To assume such a cosmic assignment, first of all it takes a burning desire to fulfill your spiritual mission and a willingness to gain all the knowledge, expertise and qualities required to become a professional speaker / teacher and a conveyer of knowledge. It is sometimes a lonely journey, and you must learn to accept rejection, criticism and judgment graciously and not to allow it to deter you from your chosen path. Without being aware of it, most often many of the trials, tests and experiences throughout your lifetime would be designed to prepare you for your ultimate profession as a conveyer of godly wisdom.

Our main focus is to bring forth each message as accurately as possible. In order to do this, we must set our linear conscious mind aside and keep the ego firmly in hand. Our integrity is of utmost importance to us, and we must always be living examples of what we teach.

Fifty years ago, if someone told you that they channeled or received messages from God, archangels or an ascended master, you would most likely have thought that they were "not quite right in the head." Today, millions of people from all walks of life are eager to read or hear the wisdom teachings which the Beings from the higher realms reveal to us. There is still much resistance to the seemingly radical teachings of the NEW AGE, however the plethora of teachings that focus on the science of spirituality, Earth science, sacred geometry and self empowerment are rapidly becoming the accepted philosophy of life for millions of people of all faiths, races and cultures. Archangel Michael tells me that the information which is being made available through dedicated messengers who are well on their personal ascension path contain the updated sacred wisdom teachings which in the past were reserved for only the most advanced disciples of the Light. We are now in the midst of a mass ascension process, and the sacred truths and laws are now being made available to everyone. We have more assistance and guidance than ever before, however we must ask for assistance, do the work necessary to integrate the teachings and then use them for the benefit of all.

I am aware that Archangel Michael's teachings through me are being translated and read by hundreds of thousands of people around the world. However, I was astounded when a friend of mine told me about a site called ALEXA.COM who records traffic data for global internet users of various websites. I was astounded to learn that the one-week average in early 2009 for my site www.RonnaStar.com was 1,507,541 hits. My overall traffic rank monthly average was 863,023 hits.

After I agreed to become a messenger for Archangel Michael, he began to gradually beam into my auric field his attributes, qualities and virtues until the time came when I had refined my frequency patterns enough for him to surround me in his auric field. This is a life changing experience and one which I will never forget. This process has taken many years for me to achieve the harmonious state that I now embody. James Tyberonn is in the midst of this process, however his transition has been and is at a much more accelerated pace. For time is of the essence, and the information he is bringing forth is greatly needed in order to assist humanity and the Earth to make this critical transition into the higher frequencies of existence. Scientists, physicists, geologists, gemologists, healers and others whose focus is on the mental and scientific aspects of spirituality / ascension are joining and even becoming leaders on the Path. People from all walks of life are being drawn to Tyb's teachings, for he is blending science and spirituality in a way that the beginner, as well as the advanced student can relate to and can understand most of the concepts he presents. The accuracy, eloquence and scope of James Tyberonn's channeled messages are astounding, and there is no doubt that his connection with AA Metatron is solid and in harmony with Divine truth.

James Tyberonn has led a very interesting and diverse life. He has been chased by hippos in Africa, rafted three times through the Grand Canyon, visited ashrams in India, served as an officer in the military, played in blues festivals as a musician, devoted tireless efforts to raising money for Cystic Fibrosis charity, coached basketball, meditated in the Kings Chamber of the Giza

Pyramid, hiked the Inca Trail, climbed Mt Fuji in Japan, spent months in the Amazon and now as a newly retired geologist, is dedicating his life to Spirit. What a transition.

I have had the privilege of watching James make this transition into the powerful role of spokesman and messenger for Archangel Metatron, and I consider him to be one of the premier teachers and conveyers of cosmic wisdom teachings on the planet. Tyb's new book *"Earth-Keeper Chronicles: METATRON SPEAKS"* is a vital source of information, channeled guidance and inspiration for us all. It is a must read for all on the path of Ascension.

I offer my eternal love and angel blessings to all my dear Soul companions. And special hugs to you, Dear Tyb, I love you dearly.

Ronna Herman, www.Ronnastar.com

I would like to offer my sincere and heartfelt gratitude to all the Earth-Keeper family who haves supported my work, attended my seminars, subscribed to my newsletter and in so doing offered so much needed encouragement in this wonderful journey. My thanks to all of you! I would like to list a few who have inspired and supported Earth-Keeper.

- Randy Monk
- Ronna Herman
- Bwaun Equa
- Carol Welhan
- Frans and Marilyn Baars
- Dr Birgit Hasler
- Alton Kamadon
- Bruce Durward
- Marilyn Springer
- Doreen Charlton
- Denise Troyer
- Dr David Mickel
- Guido Jonkers
- C.J. Walker
- Rev. Bill Buehler
- June Carr
- Bob and Sue Stinson
- Martine Racine
- Karen Bosch
- Peter Champoux
- Peter Allen
- Linda Roebuck
- Susan Phelps
- Amy Siebel
- Dr. Jaap Van Etten
- Dr. Sigrid Pichardo
- Dr. Jyoto Adrwal
- Dr. Joshua Shapiro
- Adam Yellowbird
- Israel
- Wendy-Marie Collins
- Anne Meiklejohn
- Bobby Tipton
- One Bull
- Lee Standing Bear
- Jim Self
- Rita Landry
- Ken Standing Elk Landis
- Tracey Jo and Michael Jaco
- John Dumas
- Matesha
- Dwain Briggs
- Reuben Dancing Eagle Saufkie
- Jasmine Braun
- Carol Mann
- Jinna Van Vilet

Contents

Arkansas ~ The Atlantean Crystal Awakening and the Fall of Atlantis

Archangel Metatron Channel via James Tyberonn

Greetings Masters, I am Archangel Metatron, Lord of Light!

And so again we speak of the emerging mega vortex of Arkansas, the portal of the quantum crystalline frequency. For now in 2008, this vortex vaults into global prominence with the unfolding of the dimensional lock that held the Atlantean crystalline wisdom, that of the Golden Age of Atlantis, that of the highest frequency obtained on your Earth plane.

Indeed that time has come, and in the present year will begin the unlocking of great Atlantean Crystals held for over 12,000 years in dimensional seal within the magnetic complex of Arkansas. Many of you will be drawn to take part in the unveiling, for indeed you were there as witnesses to the deluge and some of you to the consequent movement of these amazing Atlantean Master Crystals to this and other areas.

Three Atlantean Wisdom and Healing Crystals that once beamed magnificent light and celestial energy in the Temple of Healing, the Temple of One and the Temple of Knowledge on Poseida were placed in dimensional lock in the Atlantean Crystal Fields of Arkansas. The first of these will be awakened in 2008 and aligned with the other sleeping Atlantean Crystals in Brazil, Mount Shasta, Bimini and Lake Titicaca.

Indeed they will energize the Golden Sun Disc placed in Arkansas and align with the 12 patterns of same throughout the planet. Many are called, and many will come to participate in this awakening as part of an ancient contract, a sacred promise whose time has finally come. Many are there already, having been called to live within this crystalline vortex that will be fully in place in 2012. And so on the summer solstice of 2008 will the first crystal awaken.

It is not for you to activate the crystals, it is not for you to anchor the vortex. Rather it is for you to align yourselves within this completion, within this activation, within this celestial anchoring of the quantum crystalline field. Many of the Law of One are called. Masters, you have waited a long time for this; heed the call. The Master Crystals of Atlantis open the rainbow of dimensional access that accelerates your planetary and personal Ascension.

Indeed, you may access the benevolent energy and wisdom of the opening Blue Crystal of Knowledge etherically. It is not fully necessary for you to be there in physical embodiment within the Arkansas vortex for you to receive and take part in this activation. Yet there is benefit for those of you receiving the call to visit the vortex and walk the crystal fields. Those of you choosing to visit the sacred mines will be able to find a personal crystal imbued with the energy of the Golden Era of Atlantis. Indeed permission is granted, and the crystals offer themselves to you for this and myriad purposes.

Those guardians of Sirius B are even now preparing the stage and are awaiting you to join them. Send your energy; send your light and intent of highest good to what is termed the Master Blue Crystal of Knowledge. It once reigned in wisdom and light in the Temple of Knowledge, and so many of you will again see it clearly in your mind's eye. Indeed you will feel it in your heart amid tears of joy.

It is below the Earth, below the crown center of the Vortex termed Talimena Ridge in a sacred chasm built long ago. The Blue

Crystal of Knowledge stands erect, 48 feet in height and 12 feet in diameter. It will pulse and will exude a rainbow of silver, turquoise and cobalt blue as it activates in sacred contract. It holds the multidimensional wisdom and knowledge of the ages. It awaits this long destined awakening from within the sacred Ark.

The Golden Age

Dear Ones, these crystals were saved from destruction, and now you are offered the memory as to why. For such magnificent crystal beings offer all of you so much. The expansion, the sacred completion applies to ALL! It is time to remember who you were in order to become all you are in this awakening time, this return to the Quantum Crystalline Field.

You see, Atlantis existed for over 200,000 years. The vast majority of the Atlantean times were epochs of light! Only the final phase, the period from 17,500 BC to 10,500 BC, was in your terms a dark era but one rich in lesson.

We tell you in fact that the Golden Period of Atlantis was the highest level of Light Consciousness ever achieved on the Earth Plane in any advanced civilization, higher than Lemuria, higher than Mu, higher than Rama, higher than Ignacious. It has become somewhat fashionable to consider Lemuria as being the utopian civilization. And though they did achieve a relatively short lived phase of high consciousness, most in that era were not truly in physical bodies, rather in devic-like etheric states and did not face the difficulties required of the physical Earth stage. Indeed Lemuria never achieved the highly advanced level that existed in that Golden Phase of Atlantis from 40,000 BC to 18,000 BC, a time when Gods walked with men, and all knew joyful expression. It was the hallowed time when many of you walked as Star Children and later chose biology and the lesson cycles of reincarnation on the Blue Planet called Earth.

So when you consider Atlantis, DO NOT only remember its sad demise! It was but a short phase of the magnificent antediluvian world, yet it must not be forgotten. So very much is to be gained by understanding the final phase, and Dear Ones, it is time to remember.

The Atlantean Hologram

So we speak of once mighty Atlantis, and we speak to all of you. For Atlantis is a great holgramic lesson and one whose time has returned, for the memory of Atlantis is not just a healing. Indeed for some it is a requisite healing, a necessary cleansing, but for others it is also a benevolent empowerment. It recalls a time when you walked in knowledge. Regardless of your myriad roles in Atlantis, and Dear Ones, 70 percent of all on the planet at this time experienced physicality there, Atlantis now calls to you.

And the call is not just for those of you of the Law of One, it is equally called to the Aryans, to the Sons of Belial. Indeed, many of you experienced lifetimes in both ideologies. Does that surprise you?

The Deluge of 17,500 BC: Downward Spiral of the Final Phase

The time was after the second deluge, when the utopian Golden Age subsided and Atlantis split into five islands. The three major islands were known as Poseida, Aryan and Og. The two smaller islands were under the rule of Aryan and were known as Atalya and Eyre. And so in the days after the second breakup of Atlantis, the benevolent government rule by kingdom passed into the phase of a confederation, state governments in your terms that ruled each of the islands. An aristocracy of sorts evolved that consisted of two opposing ideologies. The two major components of these were the 'Law of One' based predominantly on the Isle of Poseida and 'The Sons of Belial' based on the Isle of Aryan. The island of Aryan was the most populated and exerted influence and political control over Og, Atalya and Eyre.

Poseida was the planet's most potent vortex portal complex at the time of Atlantis. The Poseida grouping and isle housed the Temple of Healing, the Temple of Sound, the Temple of One, the Temple of Regeneration and the Temple of Knowledge. Poseida housed most of the major centers for higher learning. These were placed on Poseida because of its advantageous location within the geodesic grids and because of its proximity to beneficial electromagnetic energies that spiraled upward from the Earth's core. An incredibly potent healing spring flowed on Poseida near the Temple of Healing, and its lore spawned the myths of the 'Fountain of Youth' passed down by the indigenous peoples of Florida. It flows yet into the oceans near Bimini.

The tall ones, the golden race of Atlantis of Pleiadean seed were centered in Poseida, these ones averaging 10 to 12 feet in stature. These were a gentle race of giants, involved in the cultural, artistic and educational pursuits in the heart of Atlantis. Poseida was also the headquarters and nucleus of the crystalline power grid and interdimensional tunnel system. The most advanced, complex and beautiful of the crystals were here. They were of Arcturian and Sirian construct, a living amalgam of many crystalline forms imbued with an alloy of platinum and gold. The crystals were housed in magnificent Temples, some built of marble, others of crystalline sheets of beryl, corundum and diamond. Poseida was also the capital city of the island and was called the Emerald City. The Atlanteans had perfected, using Arcturian technology, the ability to grow crystals of every structure and essence in an accelerated growth pattern within the underground crystal beds of Arkansas, Tibet and Brazil – all being Atlantean colonies accessed through the interdimensional tunnel system.

Throughout Atlantis, power grid crystals called posers were triangulated and connected through a gold-copper rod under a spherical dome that could be angled to receive specific stellar, solar and gravitational energy waves. This was the system used to power homes, offices, media, theatres, manufacturing concerns, medical centers, schools and businesses. The poser system was energized in each structure by receiver crystal units of various sizes. The poser system was also able to receive specialized refracted light and feed it into the ley energy system for use in the inter dimensional tunnel system and to feed crystalline light into conductive Earth currents to raise the frequency into a semi aware electromagnetic plasmic field of benevolent energy capable of retaining spiritual frequencies.

A great crystalline satellite, referred to as the second moon of Atlantis, floated above in the skies and received, amplified and reflected refined energy back to the crystals for various beneficial purposes. It served both the posers and the Master Crystals used in temples. Specialized energy fields were projected into agricultural and crystal growing areas to amplify and accelerate growth, as well as into universities, hospitals, schools, offices, factories and work places to create a sense of well being and invigoration for the workers and students. We will speak more of this 'second moon' later in this assay.

Now, the Temples were astonishing works of sacred geometry and stunning architecture. Many of the great temples were covered in a dome of projected crystalline amplified light, somewhat like a glowing force field. These were of various colors and glowed in day and night. These were varied according to the purpose of the temple complex. Some temples projected domes and fields of vibratory light and resonant sound frequencies that amplified the senses and chakras, others amplified the receptive learning ability of the mind and others enabled multidimensional consciousness, communication and transport. Additionally, most of the major cities during the Golden Age of Atlantis had crystalline power domes of energy fields above them.

Only after the second deluge did the capital city of Poseida have a full crystalline energy dome, and it was an incredible emerald green color. Prior to the deluge of 17,500 BC, most major cities including Meruvia, the Capital of Aryan, had such power domes over the perimeter. The one in Aryan was a soft ruby red.

In the Golden Age of Atlantis the Pyramids were three and four sided, depending on their utility and generally made of marble, granite and complex crystals. The three sided pyramids were used as antennas to draw and amplify energies which fed them into the poser grid to power homes and factories and to create energy fields for various utilities. The crystalline satellite was used to reflect the angled stellar energy waves into these triangulated grids. There were over 100 of these triangulated pyramidal grid complexes. They were set up in concentric triangulation patterns all over the planet. They established a hemispheric grid network of crystalline and electromagnetic energy, dividing the areas of Atlantis, America, Africa, Mediterranean Europe and South America into different demographic regions for powering population centers and for modulating weather patterns and tides. The areas of Mongolia and Tibet were also part of this complex, connected by inter dimensional ley tunnels. But the largest groupings of these were in the homelands of Atlantis.

The four sided pyramids were essentially Temple complexes set as full octahedrons and used for healing, learning, regeneration and spiritual purposes. These were not triangulated and generally set on hilltops or along the coastlines to receive both telluric and celestial energies. The lands below the center plane were excavated to allow for the downward pyramid of the octahedron design construct, thus connecting energetically to both above and below.

The 'Law of One' contingency of Poseida was a devoutly spiritual people who sought equality among the populace and a unification of Oneness. Although immeasurable technology and quality of life had been lost during the prior two breakups of Atlantis (the first in 58,000 BC, the second in 17,500 BC) technology still remained at an advanced level. Yet the harmony experienced in the Golden Age of Atlantis spiraled downward after the breakup into islands in 17,500 BC, as the islands grew into separately governed entities which held to very differing ideologies in the initially chaotic and difficult period of reconstruction immediately after the breakup.

Poseida and the Atla-Ra

A highly disciplined and evolved sect of scientist priests held the technical wisdom and expertise for managing the crystal power grids. The vast majority of Atla-Ra were of the tall golden race, but there were also members from the bronze, white, Lemurian brown and Cetean races. At that time there were still Cetean dolphin beings who walked on two legs, communicated verbally and breathed air in a similar fashion to humankind. The sect of scientist priests was referred to as the Atla-Ra. The Atla-Ra still maintained the highest standards of consciousness and were able to continue vibrating at very high frequencies at and above the level of 12th dimensional light and energy, remaining very pure and resonant with the true 'Oneness' concept of the Creator God. They sustained higher dimensional telepathic contact with the advanced space brethren from the Pleiades, Arcturus, Andromeda and Sirius.

The Atla-Ra Priests were highly revered and traditionally separate from, above and exempt from governmental controls. Thus they kept the knowledge and senior management of crystal technology largely in the wise and benevolent hands of the sect, although there were some technicians and crystal engineers from the Poseidan populace involved who were not in the Atla-Ra sect. The Scientist Priests of Atla-Ra were both male and female and were capable of considerable life spans, both through the regeneration of mind power and through the technology of same in the Temple of Rejuvenation. Many lived lifetimes in the same biology for 6,000 years, some to 12,000 years! Thus was the technology preserved through this holy sect. Many brilliant souls were among these. The entities you know as Galileo, Isaac Newton, Einstein, Tesla, Edison, Marcel Vogel, Ronna Herman and Da Vinci were among the Atla-Ra. Tyberonn and Oberonn were part of this group of Scientist Priests who lived very long lifetimes.

Many of you, especially your artists have great recall of the capital city of Atlantis located in the region of Poseida. It was termed the Emerald City because of the glowing green light dome projected over it. Some of your artists in fact have made renditions of the city that are quite accurate. It was an exquisite marvel of architecture, culture and engineering, by far the most beautiful metropolis that has ever existed on your planet. It was breathtaking. The city, also named Poseida as was the state, was composed of a series of concentric walls and was encircled by flowing seagreen canals. It was full of beautifully constructed temples, universities, theatres and museums. At the very center was a hill, and on top of the hill was the majestic and stunning

Temple of Poseidon, visible from everywhere in the holy emerald city. Inside the temple was a massive golden statue of the God of the Sea, showing him driving six winged horses made of extra-terrestrial platinum. The statue was embellished with precious gemstones of every color and type. The Temple was octagonal, and along each of the eight walls were built convex enclosures with platforms for amazing crystals that stood twelve feet high and glowed like transparent diamonds. Poseida had been the least damaged of the major Atlantean cities after the breakup into islands, and it was still of a very high frequency and quality of life.

Aryan and the Industrial Complex

Aryan was the largest of the islands and the most populated. Aryan was the commercial center and yielded the most influence from the economic, agricultural and military perspective. After the 2nd deluge Aryan was significantly damaged, and the infra-structure required a chaotic stage of rebuilding. In the process, the state became controlled by an elitist, affluent 'white' race, who gained control of the economy, military powers and island-state government, even though the majority of the populace were of the bronze or red race. From Aryan grew a corrupt power minded aristocracy who sought to block the 'Law of One' and utilize Atlantis' technology for control of the world. This would be done through the utilization of crystalline energy for weaponry and the use of genetic science for development and retention of an inferior race to serve as workers and soldiers.

The genetic engineering was based in Meruvia, on Aryan. It had originally been used for benevolent purposes, for seeking improved physical vehicles for those who had incarnated into physical bodies that mutated into half man half beast abomi-nations. The genetic work had been processed and developed on Aryan and was used to remove appendages, claws, feathers and reptilian skins and scales. This was done in the Temple of Purification, somewhat of a specialized medical centre. Again, we emphasize that during the Golden Age of Atlantis, its utility had been quite benevolent. Tremendous genetic advances were made, and a great understanding of cloning and the adjustment of physical limitations for betterment developed in responsible, ethical utility.

In the chaos of the reconstruction period, genetic engineering came under the control of the Belial group and then degenerated into corrupt practices of greed and power. The genetic engineering began to be used for dark ends, for creating a race of worker slaves and hybrid man animals. Just as in Nazi Germany, it was conveyed to the masses as development of a 'pure race.' In fact many of the genetic scientists were initially kept unaware of what their research and development work was being used for until it was essentially too late for them to stop it. Some of you carry great guilt to this day as a result of this work.

Law of One and Sons of Belial

This use of genetically engineered servant classes led to the great rift between the Law of One and the Sons of Belial, the latter of which became so engrossed and hardened in the materialistic ambitions of the industrial machine that they lost sight of the spiritual ethics that had prevailed over the continent of Atlantis in the Golden Era. Literally hundreds of thousands of hybrid mutations were created to work the fields, and monstrosities were brought into being with controlled minds to robotically do the biddings of their 'masters.' The agro-industrial society of Aryan became quite dependent on them. Souls became trapped in cloned bodies termed 'things' and 'others,' with genetic lobotomies and with nullified sexuality and emotional abilities. Many who were trapped in monstrous or sub intelligent androgynous human bodies still carry that horrendous pain of being imprisoned in physical embodiments that allowed for no advanced learning, spiritual growth or emotional expression.

For several millennia the two ideologies remained deadlocked over this issue, and governmental agencies were in a state of delicate impasse. The Poseidans were far too cultured and gentle in nature to attempt to fight the Aryans and sought instead to educate them and to spiritually influence them into changing their ways. The Aryans, who outnumbered the Poseidans three to one, dared not attack Poseida lest they have the crystalline energy that powered the nation shut off.

Colonial Wars

In the interim, outbreaks of war erupted between the Aryan controlled Atlantean military, as the former colonies of Atlantis located in the Mediterranean areas broke traditional ties and developed their own independent governments. The colonies, particularly those in Greece and Turkey, saw the change from the utopian Atlantis to the military fascist state controlled by Aryan, and they sought to break away. Although the Aryans had the military edge, the Mediterranean states did not succumb, and regional colonial wars raged on with neither side able to dominate the other. Factions within Aryan increasingly sought to use crystal energy to quell the adversaries. This was firmly refused by the Atla-Ra and 'Law of One' populace of Poseida.

Several futile attempts to terrorize the Poseidans were attempted by the Aryans, and each was quelled. The Poseidans retaliated by shutting down the power and negating the Aryans. The Aryans responded by stopping food supply and manufactured goods. A great impasse ensued.

The Atlantean Unification Congress

In a great deceptive scheme disguised as a plan for unification, the Aryans approached the Poseidans with the formation of a renewed National Congress to work out their growingly tense differences and to bring harmony to Atlantis. Representatives from the Law of One were sent along with the representatives from the Sons of Belial. The Congress was formalized with an equal amount of representation from the two parties. Within short order, a federation of sorts was sanctified with the promise of greater harmony. It appeared for several decades that the promise of change and betterment might come about from the National Congress. The Poseidans were encouraged, and many of them let down their guard. The Atla-Ra remained wary of the motives and sensed deception.

Initially the National Congress did improve relations, and many minor superficial laws promising unity were set into motion. Yet the primary points of opposition, the genetic slavery and crystal energy management, remained unresolved.

However, from within the Sons of Belial had risen a hypnotic charismatic leadership who seduced the masses in Atlantis into believing that they were the answer to returning Atlantis to its lost Golden Age of prominence and abundance. The leaders of this group were the souls that you know as Hitler and Himmler, indeed the supremist Nazis of your World War II. Large militaristic legions were formed, and gained the edge in covert political persuasion and power. The hybrid mutants were used to terrorize those who opposed them in Aryan and Og, and on occasion this was futilely attempted in Poseida.

Members of the powerful Aryan contingent were masters of manipulation and media. They presented well thought out arguments that masked their true intentions and promised mutual compromise. The propaganda seemed quite feasible on the surface and convinced many including moderates and some from the Law of One of its hope for harmony.

In what appeared to be a great opportunity for resolution, a law was proposed by the Belial group to bring the genetic engineering heretofore managed by the state of Aryan under federalized control and in return to bring the crystal poser system under the control of a governmental agency. A great discussion and debate occurred under much national scrutiny. A vote was set, but it failed in the congressional vote.

Betrayal

However, a compelling compromise was offered by the cunning leadership of the Belial, which allowed for the law to be passed on the basis that a council staffed with five Poseidans and four Aryans would head a new government agency for federalized Crystalline Energy control. The fact that the Law of One were given apparent majority control of the council made the plan seem very promising to the populace of Poseida. The law allowed that no changes could be made without majority consensus of the council but included the caveat that while the Atla-Ra would remain in departmental head positions, they would no longer be exempt from governmental controls. A training program was included that would allow engineers from both parties, outside the Atla-Ra, to be trained and taught the complex engineering.

Initially, despite the wariness of Atla-Ra, the system appeared to be working and bringing a better harmony.

However within two years, the wars in the Mediterranean roared back into full escalation, and the rebelling colonies seemed to be gaining the advantage. Pressure mounted to use crystal beams for war purposes under the guise of national security. A discussion, debate and vote were scheduled for the Governing Council. National security was touted, and a misaligned sense of manipulated patriotism swept the land.

Then the deception rolled into effect. To the great shock and chagrin of the Poseidans, one of the Law of One members of the Council switched positions. He was neither of the Atla-Ra nor of the Golden race. He had risen politically as a charismatic leader, a trusted negotiator who had pledged allegiances to the Law of One and gained their full confidence. He had been targeted and seduced by the Aryans and fell victim to his ambitions.

In the aftermath, this one felt great remorse and spent subsequent lifetimes attempting to make up for the mistake. For truly he did not foresee the catastrophic ending and had allowed himself to be blindly compromised under promises of power and position.

Dear Ones, such is the illusion of power. You see, when one gains power, that which may seem right can often be an illusion of ego. Each on the road of Mastery must eventually decide between power and love. Even the one you call Hitler believed that the master race scenario would enable a better future for the Earth. With one supreme physical embodiment being the single eventual race that all souls would reincarnate into, disease would be reduced and racial division eliminated by having only one 'Master' race. Even the one you call Judas in your biblical allegory thought that by placing Jesua ben Josef in detention, he would be forced to use his divine powers to reveal his Mastery to the world! Indeed the paradox is that what you term as 'Power' is often the polar opposite of love. Do you see how ego and power can deceive? Do you? Thus through 'legalized' bureaucratic authority, the use of the crystal power complex and grids became subject to governmental control under the sons of Belial and sadly could not be reversed.

The Second Moon of Atlantis

That which was known as the 'Second Moon of Atlantis,' the crystalline poser grids and fire crystals came under governmental control, and their utilization became altered as the Aryan knowledge of the programming grew. The Atla-Ra were able to defer the usage initially but in time were stymied.

Now as we have mentioned earlier, that which is termed the Second Moon of Atlantis was in fact a massive crystalline satellite. It was of Arcturian construct and was managed by the Priest Scientists of the Law of One. The crystalline satellite was an enormous unmanned sphere of brilliant engineering, approximately five miles in diameter. It had been in use since the Golden Age of Atlantis and had served myriad benevolent purposes. It amplified and controlled the various crystal beams sent from the fire, healing and energy crystals. It was somewhat of a computerized macrochip that refracted, amplified and reflected powerfully refined beams of energy for use in agriculture, weather control, tide control, healing temples, regeneration temples and the enhanced ley energy systems generated by the crystal poser system. It loomed in the skies over Atlantis and appeared as a golden 'harvest' moon, and thus was it known as the second moon of Atlantis. A rainbow kaleidoscopic energy band of antigravity plasma swirled around the sphere and often appeared as what you now term the aurora or northern lights. The satellite crystal moon did not orbit the Earth, it moved as it was programmed, self directed, constantly shifting locations in order to perform its myriad tasks over Atlantis, Africa and the eastern seaboard of Brazil.

After the Crystalline Grid complex legally came under federalized Aryan control, the Belial group integrated their own technologists into the engineering group, quickly replacing key department heads with their own. The Atla-Ra attempted to block their attempt to reprogram the satellite for war usage, explaining that overloading the satellite would dissipate the antigravity field that maneuvered it, and a catastrophic crash could occur. The Aryan scientists debunked the claim. Some of the Atla-Ra were threatened and removed, others began to mysteriously disappear. Many of the Poseidans felt intimidated and powerless, as the governing council allowed for the satellite to become a 'strategic defense weapon' and were certain that it would function as programmed to bring a quick end to the rebellion wars.

The Belial scientists with council approval reprogrammed a system bypass, and they began sending destructive thermal light beams used to initiate volcanic eruptions and massive earthquakes against colonies and nations who refused to give in to their demands. These beams were aimed at the areas of what is now Greece and Turkey and caused great devastation. Indeed it gave the zealous Aryans the battle advantage they so fervently desired, and they jubilantly increased its usage with support from most of the populace.

The Beginning of the End

The Crystal Moon began to 'overload,' weakening the antigravity field that kept it afloat. The Atla-Ra understood the implications of what would soon occur when the programming crashed, but their pleas to the council continued to be ignored.

After several months of prolonged war usage the satellite began to erratically swerve and shift, and power blackouts began to occur. Tireless attempts to correct it were unsuccessful.

The Atla-Ra were asked to lend support in correcting it. Most refused; some agreed to try to stabilize it to prevent the impending disaster. All attempts failed. The council refused the suggestion to incinerate the satellite, disbelieving that it would crash and minimizing the effects of an impact even if it did.

Relocation of the Crystals

Tyberonn and Oberonn gathered an inner group of loyalists within the Atla-Ra and Law of One to plan a circuitry disconnection and the imminent relocation of the fire and energy crystals to various 'safe' locations before the impending crash of the master satellite. This was done with the technology and assistance of those from Sirius B.

The relocation of the precious crystals was very risky, and it required careful planning and great secrecy. It had to be done before the crash of the 'second moon' and without knowledge of the governing council.

Simultaneously, other trusted members of the Atla-Ra worked furiously to hastily and safely gather data recording crystals, crystalline skulls and engraved historical records for secure placements in Yucatan, Alexandria and Giza. This was only partially accomplished, so a great deal was not able to be saved.

Atlantis had numerous power crystals located throughout the five islands and along specific relay routes of the underground labyrinth system. The Atla-Ra knew that once the modulated 'motherboard' of the crystal moon satellite lost its antigravity field, it would crash in a huge explosion. Its crashing would subsequently wreak havoc on the major and poser energy crystals, creating catastrophic secondary explosions of a nuclear class within hours or days of the crash.

The Atla-Ra and the Sirians wanted to insure that the master crystals would not be destroyed or be used for any further negative purpose and would be saved for a time when mankind could use them as they were intended. They understood that the power required to transport them would be lost after the crash and that urgent expediency was required.

Seven of the enormous primary crystals and two slightly smaller but incredible Arcturian crystals were relocated within the bulk transport systems of the underground tunnel system with help from Sirius B. Three huge primary crystals were relocated to the Atlantean crystal fields of Arkansas, two were relocated to the underground crystal farms of Brazil in the areas of Bahia and Mineas Gerais. One was relocated to the underground chasm below Mount Shasta. The great fire crystal was placed underground in a chasm below the Bimini Bank in the Sargasso Sea. The two sacred Acturian crystals were relocated in the chasms below the area of Tiajuanaco, Bolivia near Lake Titicaca.

All nine crystals were placed in dimensional locks, essentially powered down into energetic dormancy through the technology of the Sirians. Dozens of other Master Crystals were lost.

The rest, in your vernacular, is history, poignant lost history from your mainstream perspective. Truly, the paradox of the forgotten history is that it contains the most purposeful of lessons!

Indeed, after a few months of being utilized for the thermal 'death ray' technology, the great crystal satellite overloaded. Its antigravity cushion weakened, and it crashed with the accelerated velocity of a massive comet in a horrific explosion that devastated most of OG and critically weakened the tectonic stability of the Atlantean plate, vaporizing massive sections of substrata. The great crystal satellite shattered into billions of fragmented crystalline shards, which now fill the deep trenches of the Atlantic. Massive clouds of dust and smoke erupted, hiding the sun. A chain reaction of earthquakes and tsunamis devastated the island and sent enormous waves over two thirds of Aryan. Within minutes, the remaining power stations exploded with the force of nuclear bombs. The remains of one such crystal power station explosion can be seen to this day in the area of northeast Brazil called 'Sete Cidades.'

Atlantis, the eastern coast of Brazil and the western coast of Africa were devastated by subsequent earthquakes. Panic and havoc ensued for three weeks, as the remaining dry areas shook and landmasses collapsed into the sea. The land bridge that connected Poseida and Og to the Yucatan remained above water initially and was literally filled with hundreds of thousands of Atlanteans frantically attempting a horrendous panic filled exodus. Every type of seagoing vessel was filled with the surviving terrorized refugees.

And then in one shattering gasp, the remaining lands collapsed downward into the sea. The displaced seas that became known as the great flood sent dozens of enormous tsunamis spilling over South America, Africa and Europe.

It is a scene that has for many many lifetimes plagued and darkened the memories of many of you, who were indeed a part of it. Dear Ones, it is time to let it go.

Intent and Highest Good

So many of you believe that everything happens as it should happen, as it is 'meant' to be.' Dear Ones, that is not always the case. Things happen as they happen, as a set of many potentials. And indeed from the higher perspective events happen as they are 'intended' to happen, and we stress the word <u>*intent*</u>*. You are creators of your event horizons. But Dear Ones, events do not happen in the highest good until you create the highest good. The Fall of Atlantis was not the highest good, and it will not be until you recreate it as such.*

If all happened for highest good, then you would not reincarnate. In the dramatic school of hologramic 'reality,' you will cycle and recycle until you learn this. Do you understand, Dear Ones?

Time and probability are a beautiful illusionary paradox. And this is why we tell you that the Golden Age of Atlantis is actually an event horizon from your future that you pulled into your 'seeming' past!

The Crystals of Arkansas

And so now we skip forward 12,500 years, and we look at the role of the Atlantean colony you now call Arkansas. Arkansas was specifically chosen to house three major mega crystals for several reasons:

1. *It was already in use as a crystal mining and harvest area. As a result, one of the primary interdimensional tunnels from Poseida was already established, and transport complexes were available to transport crystals to the area. The Atlanteans had developed techniques that allowed for highly accelerated crystal growth, and preexisting deep caverns were present that would be quite out of the reach of mining concerns.*
2. *It was understood that the vortex in place in the area of Arkansas would play a major role in the 2012 planetary Ascension.*
3. *Arkansas' unique geology of quartz, diamonds, magnetic*

lodestone, iron, limestone and massive caverns made it the perfect incubator for the crystalline plantations. Crystals had been planted and grown in Arkansas for thousand of years before the deluge. The presence of magnetic metals in the crystal beds made it easier for the majestic Atlantean crystals to be placed in a dormant state within preexisting facilities.

4. *A very benevolent colony of the blue skinned Lemurians exists underground in Arkansas, as well as an underground base of Sirians. The two were in agreement to be caretakers for the sleeping crystals.*

Now, the three crystals placed in Arkansas are preprogrammed to a network grid that axial tonally connects to the remaining five enormous master crystals placed in Brazil, Shasta, Bimini and the two exquisite wisdom crystals placed under LakeTiticaca. These are specialized Healing, Wisdom, Energy and Transport Crystals used in the Temple of Healing, the Temple of Sound and Light, the Temple of Knowledge, the Temple of One, the Temple of Thoth, the Ruby Temple of Fire and the Temple of Regeneration.

The crystals were specifically placed in areas of supreme importance for the new Earth emerging in the 2012 Ascension. These areas would be easily accessed through the Atlantean interdimensional tunnel system and would emerge as mega vortex infinity points in your current times.

These are defined, located and placed with activation dates as follows:

Arkansas

1. *Blue Crystal of Knowledge (Interface): 8-8-8*
2. *Emerald Crystal of Healing: 9-9-9*
3. *Platinum Crystal of Communication (Bio Plasmic Interface): 11-11-11*

Bimini Bank

1. *The Ruby Fire Crystal of Energy: 12-12-12*

Brazil

1. *Gold Crystal of Healing Regeneration: 9-9-9*
2. *Violet Crystal of Sound: 10-10-10*

Mount Shasta

1. *The Crystal of Multidimensional Interface: 9-9-9*

Tiajuanaco-Lake Titicaca, Bolivia

1. *Sun-Moon Crystal of Light: 9-9-9*
2. *Crystal of Thoth: 12-12-12*

In 2008, the Blue Crystal of Wisdom and Knowledge returns. On the summer solstice, it will awaken to one quarter strength. On that which is termed the 8-8-8, the 8th of August of 2008 will the 144 Crystalline grid launch into three quarter function and in turn will activate the Blue Crystal into half power. It will trigger the time sequence program for the awakening of the others. Each will open on the triple date portals. Each will be tied to the activation of the four remaining 12-stellated facets of the 144 grid, the double penta dodecahedron. The first eight of these 12 grid-facets are now in full function.

Powering Up and Reprogramming

None will move into full strength before 2012. Only the Blue Knowledge Crystal, the Emerald Healing Crystal and Sun-Moon Crystal will be at full power by 2012; the others will take longer. They will awaken and then begin a rebooting process, followed by new downloads and reprogramming. This reprogramming will especially occur with the Bimini fire crystal, which will not be fully functional before 2020.

Their energies will be networked with the Quantum Crystalline frequency of the Ascension and tied to the energetic patterns of the Golden Sun Disc, which represents the new blueprint, the DNA in a manner of speaking for the 144 grid. The unified effect of the crystal awakening, the 144 grid completion and the completion of the 12-helix Golden Sun Disc will prepare the Earth for the heralded Ascension and for the new planet of the Quantum Crystalline field.

The effect that these crystals have will initially be quite subtle, yet that subtle strength will be immediately discernable to empaths. The effect will be a greater sense of well being, especially within the vortexes of Arkansas, Lake Titicaca and Bahia, Brazil. Each who visits these vortex arcs of the crystals, whether in physical or etherical phase, will be offered the opportunity of empowerment with the knowledge and potency of their crystalline codes. Yet be aware that you must be of a sufficient light quotient to absorb this frequency.

The first phase represents wisdom, and the second phase will be the etheric healing of Atlantis and the tie in to the new Golden Age of humankind. There will be a golden rhombus formed between Brazil, Lake Titicaca, Arkansas and Mount Shasta, and in 2020 this will become a five-pointed star that forms a global network with all infinity points on the planet.

The awakening of the crystals beginning in Arkansas will open the gate for the receptive dimension that marks the return of the star brethren. Return is a misnomer, as truly they have never left, rather mankind lost the frequencial ability to perceive them.

Closing

The Atlantean Crystals open the etheric gate, the dimensional portals that will allow mankind of carbon based life to interface with those of bioplasmic life. You must understand, Masters, that you exist in both forms within multidimensional parallel. Indeed you are the Sirians, you are the Pleiadeans, you are the Arcturians and much more! You are all the multidimensional sparks of the Master Creator God! It is true! The gate will open in 2008 in the crystalline vortex of Arkansas, and those of you who connect to the blue crystal will be able to experience your own multidimensionality quite joyously, quite easily in linear time as you develop the light quotient and impeccability to do so. We do not imply that it comes to you without work and will, but it opens to you in greater facility than you have known for eons.

Dear Ones, within the dynamic of the Arkansas Vortex lies a crystalline energy like none you have experienced since that Golden Age of Atlantis 20,000 years ago. Within the quantum crystalline frequency can you heal that which must be healed, open that which must expand and offer an extraordinary alignment to the knowledge of exquisite bliss. Within this rising energy will you be able to connect with a living crystalline force that will assist your merkabic expansion into your multidimensional divinity. Into the Divine Self! It is true, Dear ones, and that time is NOW.

Masters, it is the end of the beginning! The dreamer awakes! I am Metatron, and you are dearly Beloved!

… and so it is. AA Metatron via James Tyberonn

The Magnificent Om Crystal of Shasta

Archangel Metatron Channel via James Tyberonn

Greetings beloved! I am Metatron, Lord of Light.

And so again we speak of Shasta. Indeed, it's importance is tantamount to the expansion of the Crystalline Field.

Now, far beneath the expanded massif of the mega portal that is Shasta stands an amazing crystal of Atlantis, a crystal of Arcturian origin. It glows in a silveresque hue of argentum, emitting the frequency of crystalline platinum and etheric gold. It once stood in the great Temple of One on the Atlantean Isle of Poseida and was formally called the Crystal of OM. The Atlantean scientist-priests, the Atla-Ra, referred to it as the Crystal of Multidimensionality. Indeed, this sentient crystal is omnipotent in its frequency and holds within its structural alloy of platinum, gold and hyper quartz the ability to fold, concentrate and amplify that which you term light, space and time.

The Platinum Ray

The OM Crystal draws to itself and emits what is termed the Platinum Ray, an omnipotent energetic beam of multidimensional resonance. The Platinum Ray is essential to the balance of the coming Ascension. It holds within it what may be termed the code of Universal Law, of the Cosmic Axiom. It holds within its frequency Universal Truth, multidimensional Truth that expands and migrates into every aspect of the Cosmos. It has the nurturing feel of the femine frequency, yet it is in fact the optimal balance of both the female and male vibrancy. The Platinum Ray is one of unity and balance that in fact lessens the differential between left and right aspects of the brain. It unifies consciousness and helps create a great synchronicity of mind within duality.

The magnificent Crystal of OM is one of nine enormous and potent Temple crystals that were saved from the deluge of Atlantis before the fall and crash of the crystalline spherical satellite, termed the second moon of Atlantis. The Crystal of OM was and is the most complex of all of the Atlantean crystals, second only to the great crystalline satellite in its omnipotent capabilities and technical intricacy. Indeed it was the most prized and first of the magnicient Temple crystals to be sought and relocated in the chaotic final days of Atlantis.

The Crystal of OM

The crystal masters of Arcturus constructed and fashioned the OM Crystal in an alchemical process that involves a liquid conglomerate of synthesized quartz and diamond, termed hyper quartz, amalgamated into a silvery clear alloy with crystalline platinum, gold and mercury. The living composite was fused with life force and manifested into a stunning geometric form, whose crystalline structure encompassed the golden spiral of all within the Metatronic Cosmos. Accordingly it is in harmonic oscillation with every dimension, every universe and every parallel in the Cosmos. It operates both in carbon based life systems and in bio plasmic systems, tying together and existing as both matter and antimatter. It has 144 facets fashioned in equitable concave and convex indentures around the circumference of its cylindrical length. Its dual points are angled at precisely 51.8 degrees. Indeed this specific angle is significant and was utilized in the construct of the Great Pyramid. This pyramid indeed represents the frequency of the Earth plane intended to be a representation of the spherical Earth, the height corresponding to the radius joining the center of the Earth to the North Pole and the perimeter corresponding to the Earth's circumference at the Equator.

The number 144 is a very complex frequency as well. It represents the code within the golden spiral of all the geometrical base units, termed atoms, in your universe. Your universal matrix contains within it a living geometric constant that has always been. It expands and contracts both in matter and antimatter in regular measures. The very geometry produces a distinct tonal quality which is tuned by the mathmatical proportions of all celestial spheres.

The Golden Phi

The golden ratio is a constant ratio that offers the key to the Cosmos. It contains within it the vibratory resonance of all matter and indeed propels the doorway into parallel dimensions and planes of antimatter. Your own scientists have proven, seen and experienced that the developing in a fractal manner of the golden phi spiral is a geometric pattern that is repeated at every scale, from a cell to a star. Simply put, the golden ratio is where the ratio of the small to the large is equal to the ratio of the large to the whole. So it is in your universe; as above so below applies. It offers many doorways that have not yet been explored by your current level of knowledge.

The great Crystal of **OM** *beneath the matrix portal of Shasta incorporates the golden mean and the universal frequency. It does so by the resonance of its intricate faceting and angular design. It exists in the now and is contained within that which is termed the past and the future. Can you imagine that? It does so to a large degree, from your reference, by its utilization of the golden mean and its frequencial pulse.*

Now, the Earth plane, your planet as you know it is a specific focal vector for consciousness. At your level you think your planet is divided into areas of land and water, continents and oceans.

The awareness faculties of humankind, indeed your consciousness, are tuned into frequencies of perception that give you that impression. But from a Metatronic perspective, your planet is also faceted into frameworks of time and of probability areas termed parallels. So all time, all parallels coexist in greater reality. As such all of your epochs, all of your time eras, all of your many civilizations coexist at once. Indeed, from this extraordinary panorama time is simultaneous, so all eras of Earth history are embedded in time and clearly exist in the eternal NOW along with your current time vector. Within the golden mean, within that which you term portals there are certain bleed throughs, certain apertures that allow for movement. The **OM** *Crystal is capable of tapping into these and harnessing their utility.*

In the truest sense your now, your present moment is a psychological platform. This is true whether you can grasp the concept or not, and we realize that it is an enormous leap of understanding. Accept it as an axiom, and indeed so doing will assist you in managing the understanding of your true multidimensional nature.

The Crystalline Network

So clearly was the OM Crystal was the most complex in its construct and utility of all the magnificent Temple crystals in Atlantis. This is not to say that the other Atlantean crystals are not of enormous capability, complexity and purpose, indeed they are. We have told you that nine utilitarian Atlantean Temple crystals, stunning specimens of Arcturian and Sirian construct, were transported through the hyper dimensional capacity of the inner Earth tunnel system and relocated to specifically chosen areas. We have also told you that the Blue Crystal of Knowledge, located in the Crystal Vortex of Arkansas, would be the first activated. Indeed it has been rebooted and surged into the first tier step up to approximately 25 percent power on the June solstice of 2008. We tell you that on the frequencial triple date portal of the 8-8-8 that it will launch into half strength and in so doing will also trigger the 144 Crystalline grid into two thirds functionality. The 8-8-8 activation of the Blue Crystal will trigger a contractual surge of axial tonal energy networking to the other eight crystals. Four of these are scheduled to come on line on the 9-9-9, September 9th 2009. These are the OM Crystal of Multidimensionality in Shasta, the Emerald Crystal of Healing in Arkansas, the Golden Crystal of Regeneration in Brazil, and the Sun-Moon Crystal at Lake Titicaca.

Indeed, that which may be termed as a Crystalline Network is tuned to the activation of the 144 grid and correlates with the function of that termed the Golden Sun Disc:

Arkansas

1. *Blue Crystal of Knowledge: 8-8-8*
2. *Emerald Crystal of Healing: 9-9-9*
3. *Platinum Crystal of Communication (Bio Plasmic Interface): 11-11-11*

Bimini Bank

1. *The Ruby Fire Crystal of Energy: 12-12-12*

Brazil

1. *Gold Crystal of Healing Regeneration (Mineas Gerais): 9-9-9*
2. *Violet Crystal of Sound (Bahia): 10-10-10*

Mount Shasta

1. *The* **OM** *Crystal of Multidimensionality: 9-9-9*

Tiajuanaco-Lake Titicaca, Bolivia

1. *Sun-Moon Crystal of Light: 9-9-9*

2. *Crystal of Thoth: 12-12-12*

The Atlantean Temple of ONE

Near the Temple of ONE was what may be termed a center for advanced learnings, an exclusive university of sorts for the select of the initiates of the scientist-priests. It was called Atla-Ruum. Ruum is a word from the sacred frequencial semantic of the Atlantean Golden race, a term meaning 'Mystery Initiate School.' The tonal qualities of this term will open memories for many of you who were among the Poseidans, especially those of you who were among the Atla-Ra. This wisdom school taught advanced mathematics, physics, sacred geometry, hyper dimensional geometry, crystalline chemistry, alchemistry and energy dynamics as well as the disciplines of astral travel and achievement of multidimensionality. Most of the faculty were of the Golden Race yet included Ascended Masters such as Thoth and Masters from the Pleiades, Arcturus, Andromeda and Sirius A and B. Those of the Golden Race were tall beings, ranging in height from 9 to 12 feet by today's measurement. They were direct descendants of the Pleiadean Masters and were very gentle, peaceful beings of an extremely high intelligence and resonance.

The Temple of ONE was itself located on a granitic mound at the northern portion of Poseida in the sanctuary of the scientist-priests, the Atla-Ra. The Temple of ONE was built to a very unique dynamic of sacred geometry, and the location was carefully and specifically chosen for its natural energy field and celestial grid alignments. The OM Crystal was situated in the center of the amazing Temple. It did not sit on a base, indeed both ends of the OM Crystal are angled into points. It was held in suspension within a series of spinning rings that spherically surrounded it. These were of a metallic alloy which condensed an antigravity field that suspended the crystal. The entire apparatus could be angled for specific celestial alignments. The crystal itself attracted and emitted the antigravitational plasma field. Sound, pure tones were used to both activate desired aspects of the crystal as well as to calibrate the requisite density of the plasma. Differing tones and their correlating plasmic densities created specific oscillating colors within the antigravitational field. The crystal itself glowed in a silver hue that at certain angles and intensities changed to a white platinum violet shade.

Base Twelve Mathematics

While the cylindrical body of the crystal has 144 alternating convex and concave facets, the two points are different. The base point has 1,728 facets and the upper point 12 facets. Thus the sequence of 12, 12 squared and 12 to the third power, as Atlantis and the Atla-Ra used the base 12 system of mathematics.

It may surprise you to know that the conversion to base ten mathematics which occurred in most societies after the fall of Atlantis was chosen because humans count with their ten fingers! The exceptions were the Sumerians and the Mayans who used base 60 and base 20 respectively. It is interesting to note that humans still measure time within the duodecimal remnants of the Atlantean Base Twelve. Do you not count hours in phases of 12, as well as the months of the year?

We will tell you that in the future, mankind will change the base system of mathematics to more closely align to the changing cycles of time. We have told you previously that time sequence is speeding up due to the altered ratio of inner core counter-spin to the Earth's rotation. The time unit cycles occur more rapidly now and are no longer in alignment. Your cycles are faster, but your measurement has not shifted as yet.

Now, we have spoken about time cycle acceleration and the need to convert to a base twelve duodecimal mathematical system. The channel has queried as to the current validity of the triple number dates, which we have told you before are numerical portals that activate the major dodecahedronal facets of the 144 grid.

Activation of the 144 Crystalline Grid

Before we answer, let us review the activation process of the 144 grid. The grid is a complex geometry. The double penta dodecahedron by specific design includes 144 pentagonal and triangular facets in a symmetrical crystalline matrix. The double penta dodecahedron is stellated, which means that each of the major 12 facets are raised in the center to form five sided pyramids. Thus there are 12 pentagonal pyramids. Each facet is counted both on the base and on the pyramidally raised portion. So if you consider that each pentagonal surface contains five isosceles triangles and one pentagon, you have 72 facets. Additionally, each of the 12 stellated aspects have five triangles and one five sided pyramid. Thus you combine the surface and stellated portions to achieve the Master Number frequencial of the twelve squared 144. Do you understand?

Now, the aspects of the triple dates are quite interesting. After the millennia, there is a unique sequence of dates that take place between 2001 and 2012. The triple dates are numerical frequencies that uniquely occur in this abbreviated format. They occur sequentially for twelve years as:

- *1 January 2001, 1-1-1*

- *2 February 2002, 2-2-2*
- *3 March 2003, 3-3-3*
- *4 April, 2004, 4-4-4*
- *5 May, 2005, 5-5-5*
- *6 June 2006, 6-6-6*
- *7 July, 2007, 7-7-7*
- *8 August, 2008, 8-8-8*
- *9 September, 2009, 9-9-9*
- *10 October 2010, 10-10-10*
- *11 November, 2011, 11-11-11*
- *12 December 2012, 12-12-12*

The grid significance is that the frequency of each triple date portal is designed to be the 'activation trigger' of one of the 12 major dodecahedrons' surface base plates of the grid. Accordingly, this axiom infers that the 144 Crystalline grid is launched into functionality by one twelfth each year from 1 January 2001 up through 12 December 2012 when it achieves full programming input. In the interim between 13 December and 20 December 2012, it integrates full programming and reboots into full launch on the winter solstice of 21 December 2012, the heralded and prophesized Ascension. In aspect, a key piece of what takes place on the triple frequencial portal dates is around the 'human grid' alignment to the 144 Crystalline grid. That is why these dates are being recognized by empaths as important dates for spiritual alignment gatherings across the planet. Indeed it is not specific astrological alignments that specialize the triple date portals, rather it is the vibrancy of the numbers themselves that correlate to the 144 grid. This kinetic force is also impelling the renewed 12 strand helix of the Sun Disc to furbish a design schematic into the 144 grid that provides what may be termed the DNA programming of the pentagonal facets of the double penta dodecahedron.

Now, we have not forgotten the question raised by the channel as to the current validity of the triple number dates, when taking into account the acceleration of time cycles and the need for a base 12 duodecimal measurement of mathematics. The answer is this: your measurement of the annual solar return is already in an approximate base 12, twelve month system. And the count of the triple dates is a 12 count, there are twelve of them. So in brief, yes the triple date portals are indeed valid.

The Speed of Light is Increasing

But we will add the caveat that the full throttling of the 144 grid itself on December 21, 2012 will in fact increase the disparity, the difference between the quickening time unit cycles and your measurement of them. For not only will the new grid upon its completion accent the time sequence ratio between the inner core spin and the outer rotational spin of the planet, it will also enable a greater velocity of light to bath the planet. Light will increase from its current approximated speed of 186,000 mps to well over 210,000 mps.

Part of this light acceleration will be contributed by the activation of the Atlantean crystals. Particularly will the OM Crystal of Shasta and the Sun-Moon Crystal of Light of Titicaca effect light velocity. Both of these are activated in 2009. When all of these crystals are activated and networked in 2012, the effect will be prolific. However this cycle will not reach its pinnacle until your year of 2020, because certain of the crystals activated in the twelfth phase of the 12-12-12 specifically the Fire Crystal of Bimini will require up to eight years to fully activate.

Now, we will add that while there were nine omnipotent crystals relocated from Atlantis, there are others that were already in specific locations and did not require movement. The major of these are located as follows:

- *Tibet*

- *Lake Baikal, Russia*
- *Roslyn, Scotland*
- *Araphat, Turkey*
- *Giza*
- *Montserrat, Spain*
- *Sri Lanka*
- *Bali*
- *South Pole*
- *North Pole*
- *Hawaii*
- *Japan*

There are secondary, what may be termed satellite crystals in many other locations including Tasmania, New Zealand, Glastonbury, Canada, Tanzania, France, Greece, Capetown, Chile, Easter Island, Moscow and Greenland, to name a few.

Activation of the Crystal of OM

In a manner of speaking, these great crystals are akin to your computer systems, of course exponentially far more complex and potent, but there are certain similarities in their function. The great crystals will each go through something similar to a re-programming and rebooting. This process of programming and reboot is specific to the increase of frequency and dimensional access of the Earth matrix. It is not effecting the aspects of the crystals and their higher networking that exist in other facets of multidimensionality, you see. In a sense what is occurring is a reformatting for the changing Earth matrix.

The new moon of August 30, 2008 will begin the formatting and programming of the OM Crystal. The process will take just over 12 months for this complex crystal. It will go into what may be termed a warm-up phase, an idling stage with specific programming during key astrological periods including September 10-13 2008, the equinoxes of September 21 2008, March 21 2009, June 21 2009 and the full lunar eclipse in July 2009. It will launch into 25 percent power on the triple date portal of 9-9-9.

From the dates of September 10-13, 2008 there are a cluster of five astrological septiles, the most important being Saturn-Chiron. This is an important Portal Connection, and Saturn facilitates materialization. Its significance to the **OM** *Crystal is that the septile angle is 51 degrees, exactly as the angle of the great pyramids and the* **OM** *Crystal itself. During these warm-up phases prior to activation, it is well for humankind to direct their alignments into the* **OM** *Crystal. It is not for mankind to align or anchor the crystal per se, rather to align themselves to the energy of the crystal through the process of visualization and the sending of the vibration of love.*

Tapping into the Crystal of OM

Now, many of you may ask what is the effect that the OM Crystal, indeed the Ascension itself, will have on me? How will the change from the 3rd to the 5th dimension effect me? The effect will be one of a sense of unity. There will not be a tangible sense of differentiation between the planar frequencies of the 3rd and the 5th dimensions. It will rather be a blending. You will have a greater sense of well being, a feeling of being more supported and indeed of supportiveness within the self. By 'tapping' into the OM Crystal, you will feel a greater sense of balance; you will feel more joined to aspects of your multidimensionality and indeed within physicality. The mind will access more freely its own compartmental aspects as the dimensions become melded. There will be less of doubt, less of feeling lost and fragmented, **less** *of the internal questioning of what is right and what is not.*

The mind will become unified, more integral. There will be less polarity per se within duality. But it does not mean no duality, no polarity. Those that say the Ascension will end wars, will end hunger, will end fighting and greed, no that is not the case. But Dear Ones, all of those will lessen to visible degrees; the arc swing of duality extremes will be shorter. The pendulous movement between extremes will somewhat vacillate. Choices will be clearer, pathways more succinct. The way of impeccability for each of you who chooses to tap into the OM, to actively and fervently seek your own Ascension, will be more obvious to you than it is currently. In a sense a new paradigm, a new system of Knowingness will become available to you and will offer itself to all who seek it.

The effect then is that you will find it easier to be fully present within the NOW. And always the NOW, the eternal moment of the present NOW is the fulcrum of power. It may seem to be paradoxical from a 3rd dimensional perspective that by merging the many facets of multidimensional aspect, one is afforded greater focus within duality. But it is so. It is so, because it is the wholeness of multidimensionality that brings the integral nature of your true self. And within that wholeness is one's true causal self, and therein Universal Truth exists more clearly. Within the Now of multidimensional time and geometric consciousness you are more creative. You are more empowered and less effected – in that geometric field you are not effected by untoward issues of stress, of problems. You are in the field of solution, of resolution, you see. You are at the control panel of your experiences.

Is there work to be done? Yes, of course. But tuning into the OM, into the higher dimensions, into multidimensionality offers the tools of mastery. Many of you have the notion that the soul is at 'rest' on the other side of that which you term the veil. Your epitaphs are inscribed with 'Rest in Peace.' Dear Ones, there is peace, but it comes through work. Ascension to the celestial realm, Jacob's Ladder in your metaphor, is not static! Dear Ones, you are either moving up or you are sliding down; it is one or the other. By design, consciousness swims upstream, otherwise it gets moved downstream by the current. Consciousness is not stationary! And so we offer no hope for the lethargic. So you see, your pillowed concept of eternal rest is somewhat erroneous. However through traveling within yourself, you will discover the unity of your consciousness with other consciousnesses. You will discover that multidimensional love and energy charges consciousness with dynamic exquisite energy and movement to all things. This will invigorate the SELF; it will not lead you to want to sleep on the proverbial marbled steps of Heavens Gate. It will instead inspire you to take a better hand in the job of creation. For indeed there are 'energy packets' that offer extraordinary impetus to you as you ascend to each new level.

Consciousness is ever expanding, and so at every level there is still another one higher to ascend into. And while from your 3rd dimensional stance within duality this may appear burdensome, it is not. That could not be further from the truth. Growth in consciousness is in fact embellished in vital streams of vigorous joy. Such is the nature of consciousness; such is the complexion and disposition of 'First Cause.' Indeed there are points of review, points of recharging, but these are not stagnant. Indeed there is great activity and profound validity within them.

Dear Ones, the great Crystal of OM, of Multidimensionality offers an incredible energy that once launched on the 9-9-9 will have myriad effects on the 144 Crystalline grid and on the Ascension itself. The ability to 'tap in' to its energy will depend to a large degree on the light quotient of each individual. Its initial effects will be subtle and graduate each year. It will indeed be in full motion by the 12-12-12.

The Mega Portal of Shasta

Now the greater massif, the greater parameter of the Shasta mega vortex-portal complex is capable of folding space and time within its field. It condenses and 'zips' space more potently than any other area in North America. Its capacity for doing so will expand as the OM Crystal goes into activation.

You see, currently a cubic measure of space in the upper meadows of Mount Shasta holds upward of 36 dimensional frequencies, at certain times as many as 44. It is the blend of folded energy varied by celestial alignments and terrestrial magnetics that trigger the variation. Shasta is far more active in some periods than in others for this reason. Dimensional gates open and narrow accordingly. When one learns through intent and determination to increase vibrational frequency into the merkabic realm of crystal resonance, then the doorway is open, or accessible in your terms. Do you see? Not everyone who visits Shasta will access these portals and doorways, but all those who are capable of increasing their vibratory rate to that of the crystalline frequency will indeed. It has always been such in infinity points, stargates if you will.

Now, Mt Shasta anchors a very complex multidimensional overfold and does so through its very conscious geometric grid, complex frequencies and unique electromagnetic field. These allow light portals from many higher dimensions to coexist within the Earth dimensions precisely in the massif of Mount Shasta. Among those are the ones from Sirius B, the Pleiades and Arcturus as well as many others, and we repeat, many others. That is correct. Mount Shasta is quite prolific in the higher dimensional communities that are accessible to seekers, you see.

Now Mount Shasta, as all with all infinity points upon the planet, has a geometric projection that defines to a degree its nature and purpose. The overlay of Mount Shasta is an octahedron, the double pyramid: "As Above so Below." This is the geometric crystallization found in many of your diamonds. Understanding this can assist you in the experience you seek at Shasta.

So you see, if you seek to communicate with the infinite nature of sacred sites, then it is suggested that you make an effort to understand the multidimensional, geometric and geological aspects of them. You must understand and accept that each infinity point is aware of itself, fully conscious in your terms. Your indigenous people always asked permission to enter such places, sang songs of prayer and made offerings. And Dear Ones, we tell you it was appropriate. The Earth will respond. The Spirit of Place will open itself to you. Now truly, the Living Earth does not necessarily need the offering you bring in a material sense. But the very spirit of giving you thus demonstrate is one that opens you and creates the resonance of the gifts that the higher Consciousness of Place will return to you in kind. Do you understand? Make it known why you are here and what you seek, and you will indeed feel a tingling of energy rise up your spine, and a direct connection will occur.

The resulting experiences will open to you in ways that are beyond perhaps what you expect. Trust them. They are possibly far more revealing, far more valid than you even imagine. Bring your greater awareness to the forefront; quiet the questioning mind.

The channel has spoken in his articles of many incredible experiences and visions he experienced within the energy of Mount Shasta. We ask the channel, "Do you doubt your experiences?" These were indeed quite real. We tell you that dreams and visions are indeed actual; they are real and often the catalystic means for dimensional travel. Do not doubt them, rather learn to navigate within them.

You see, most experiences in such dimensional doorways at Mount Shasta will occur in one of two ways. Either one passes thru the doorway that reveals itself into other dimensions, to see and to feel what is on the other side of the gateway ... or one simply waits at the threshold for manifestations of 'the other side' to appear. The former seems to be a vision held in the mental eye, the latter an incredible manifestation of a Teacher or Master into what you would term the physical world albeit for a brief period. In both cases, the initial tendency among humans is to dismiss the valid reality of what occurs.

Dear Ones, you must allow the expanded paradigm to exist as more than simply a possibility. You must allow it as a valid expression of multidimensional reality. As REAL! There are far more parts to you, Dear Human, than you truly realize, and upon the Earth Plane you are far far more than the sum total of your parts! Allowing yourself to expand is only a piece of what Shasta gifts to you, Dear Seeker. The other part is up to you, and that is to accept and believe what has occurred. When Mount Shasta offers you an experience that is far removed from your daily accepted reality, you are simply being shown a piece of the greater reality. It is understandable that you might initially doubt it, but you were shown the greater truth because you expanded, through greater effort than you may realize, to the point where you could perceive it. What does that tell you? It tells you that within your expanded perception lies the field of all probability, of all possibility. Some you inhabit and some you do not. Some of the realities have never known biological life as you know it. Some contain more advanced life beings, etheric in nature but not physical yet human in heart. There are no regimented rules that will limit the potential of such travel other than the blockages you would create for yourself in nonacceptance. In truth, many of these realities are closer to your true home than is your physical **experience** *upon the Earth. That is why the removal of fear is an essential graduation to the higher dimensional reality validity of what one experiences in this very real system of growth potential.*

Now, we will add that it is often the case when humans first step into this wider reality that they temporarily or permanently can lose their 'grounding' on the Earth Plane. To remain grounded, the human must learn to manage the integrity of the Human EMF (electromagnetic field). One must learn how to open and to close the awareness centers, the chakras. Was this not the lesson of the Lemurians? Having an out of body **experience** *may not be appropriate when one is being chased by a tiger or when sitting in an office doing accounting work! Learn to manage the system, as you are on the Earth for certain valid purposes of the Earth and through discernment utilize your ability for chakric, merkabic travel when it is appropriate to do so. Do you understand? Do not become so ungrounded that you are not capable of operating within physical dimensions. There are times when it is appropriate to have both feet solidly on the ground, even in this Ascending New Age of the Earth.*

Now, the channel asks of the origin and significance of the energy body of Mount Shasta. We confirm that he, Mount Shasta, is Lemurian but also Atlantean. And although the volcano itself is somewhat dormant at this time in terms of what you consider volcanic activity, in dimensional terms it is very expansive, very active. Currently Shasta is in commissioned transition, in the task of balancing energy harmonics that exist between Atlantis and Lemuria. And for that reason it is a Lemurian area and yet an Atlantean masculine volcano, twinned with parity to the femine.

Do you understand? It is significant for both Atlantis and Lemuria. You see, Mount Shasta transfers energetic excesses from one to the other, seeking equilibrium within integral dimension. That is what creates such dynamic charge in this area. It is why one becomes dimensionally enabled and energized upon that mountain, you see?

Now, while the lands of Lemuria and Atlantis coexisted for over 800,000 years in your terms of time measurement, their civilizations coexisted only some 50,000 years. One projected heart, the other mind. One was excessively male, one excessively female in a manner of speaking. Can you see why these must be brought into BALANCED alignment for the Ascension? Do you understand that in a very valid sense all time on your planet coexists in multidimension? So we tell you that Mount Shasta is in a special role to blend these, to attune these in multidimensionality, and accordingly Shasta is very specifically in harmonic oscillation with your islands of Hawaii and Mount Fuji in the Pacific areas, and with Santorini in the Mediterranean and the Canary Islands in the Atlantic. The former hold aspects of Lemuria, the latter of Atlantis. Do you understand why it is volcanoes, specific ones that are capable of such commissioned undertaking on your planet? We tell you that it is because of the dimensional expansion capabilities of their electromagnetic fields.

Now, it is not every volcano that is so endowed. We will say that this varies according to their placement, geomagnetic properties and consciousness embellishment. There must of course first be the correct frequencial potential of sufficient raw energy there, and it must be of a specific blend. Now, as the channel has an affinity for gems, we use this analogy. One can find a rough diamond that appears to be a mere carbon deposit. But if it is burnished and proves to contain clarity and symmetry, it can be polished and faceted into a gem of luminous beauty that refracts light and energy in wondrous ways. As such do volcanoes and other sourced power points evolve into the awareness of dimensionally expanded sacred sites. But the energy potential was there first.

Now indeed, when that which you might refer to as a great historic event occurs in a field or valley or mountain, and multitudes of people are thereafter drawn there to visit the location of the great event, the energy they amass can also polish the location to a higher frequency, and a sacred site of sorts will evolve by amplification. But it is not just the raw telluric energy, not just the geometry. It must also be embellished with the sacred. And that is so misunderstood. Your mainstream physicists, mathematicians and scientists currently have many concepts, many systems of thought, of belief, of paradigm that are being applied to physics as simply a complex mathematical challenge. But they omit the sacred. The two must be coupled for humankind to truly understand the nature of the living Earth and consciousness of the Cosmos.

Now, many have experienced the inner dimensional etheric city located within Mount Shasta that is referred to as Telos. Dear Ones, we tell you it is indeed real. It is inhabited by an advanced cycle of those you term Lemurians, the blue skinned ones who have obtained full consciousness. Some of your explorers and writers have experienced this inner world and referred to it as Agartha. These beings originally escaped into the inner hollow chasms of your planet some 25,000 years ago, in your measurement. Although assigning a truly accurate time line is difficult, because their time sequence is not aligned to your own.

So indeed, they were assisted in the process. Your inner Earth was in fact populated long before you imagine. We tell you that there are beings of Divine Intelligence that you would consider 'extraterrestrial' that in fact can quite truthfully claim the Earth as their own, more so than humanity. Now, the inner Earth beings of Lemurian descent, as we have told you in other assays, are far more highly evolved in spirit than is humanity at this time. We have told you that their bodies are adapted, changed to their new environs. The inner Earth beings are less dense but are indeed physical. Their skin has taken on the hue of green and in some cases a green blue for several reasons. The water they drink is highly mineral in content containing oxidized copper and other metals in a higher concentration that tints their bodies in a copper blue patina. Also, the light they absorb is an ultraviolet blue such that at times they would appear to fluoresce in that hue. These beings are wise and have found their unified centre. They are peaceful, espousing no dogma, no religion other than the love of source and an understanding of what is termed nirvana. They exude great tranquility. They are indeed quite aware of you humans on the surface but have no desire, no responsibility to intermingle with humanity. Why, you may ask? Because of several reasons, the most prevalent being that they are aware of your violent nature, your fear and your lack of mass spiritual development. They are aware of your physical diseases that could infect them, and they are quite unable to sustain your solar rays. In their own reality, in their own linear sequence, and indeed it is a different time ratio and continuum than your own, these ones have evolved and set a unique course that is very near its own graduation, its unique completion. Their environs are sustained by a crystalline magnetic force and by a light source that is emitted from your planetary core, and their bodies have morphed into fitting within its unique dimensional paradigm. So their interaction with humanity is quite limited. In truth they are in a parallel from your own reality and world and only have brief and limited areas of dimensional interface with your time ratio. Their mental capacities have adapted to emit an internal light into the external, and as such the chasms they create are as bountiful in beauty and light, etheric light as the most beautiful of your mountain valleys on the exterior.

In dimensional terms, the Agarthan city of Telos has its existence both in the physical and in the etherical realms. In the physical, it exists just above the level of the 3rd dimension and measures some 20 miles in length, three to five miles in width, with a ceiling of approximately two miles. Thus has it been experienced by many of your adepts in human biology through the mechanism we will refer to as 'lucid dreaming.' We will add that such experience is by invitation only.

There has indeed been contact between the inhabitants of the inner world and your governments but not an intermingling. There have been messages of the need for the human race on the surface to be aware of the coming changes. Those of the inner world are primarily on a set course, a different cycle, and one that is close to completion.

Now, above the dimension of the 4th exists the light city above Mount Shasta. Within this realm are many of those you refer to as the Ascended Masters and certain of the benevolent extraterrestrial beings such as the Pleiadeans and those of Sirius B. Now, would it surprise you to know that many of you have coexisting lives within all of these realms? Would it surprise and overwhelm you to know that many of you are extensions of omnipotent consciousnesses that you call the Ascended Masters? Your biology in its dual nature separates you from this knowledge until you learn to operate the 'back brain' or subconscious nature of your being. Has this not been the way taught by your Buddhist and other adepts for millennia, to quiet the conscious mind and thus find your true nature? Those of you that have chosen to evolve with the planet in human biology have done so on an exercise of choice, to fully grow and master the experience of Planet Earth. You are much closer to do so than you truly realize. And when you pass into higher dimension, you understand that a physical existence upon the Earth is just a heartbeat of time from the higher realm.

Beloved Saint Germain has chosen Mount Shasta as one of his homes of interface with your planet. His presence is easily accessed upon the peaks and in the Valley of Upper Panther Meadows and surrounding slopes. He is indeed an Ascended Master and one who can be termed as 'modern' in many of his expressions. Modern in that he has had a variety of lifetimes manifested on your planet within the past centuries, some known as such and others not. The Ascended Masters are a group dedicated to assisting humankind achieve Mastery on the Earth and have anchored light cities in specific points on the Earth Plane for the dissemination of their energies. Mount Shasta is one such point. As we have said in the beginning of this communication, it is the magnetic nature of Mount Shasta that provides the raw template to enable the dimensional fold for higher dimensional influx into your terrestrial template.

The channel has asked if it is necessary to physically be in such infinity points on your planet to access these Masters. We tell you that it is not. The most potent infinity point you carry is within your heart, Dear Ones. But let us say that access can be more easily obtained within these points.

So you see, what is made available to you on Mount Shasta is a glimpse into the infinity of who you really are. Shasta opens the dimensions as a peacock opens its fan of blue green feathers and for an eternal moment shows you the depth of your true beingness ... at the time you are ready to see it. And Dear Ones, it takes desire, desire and work to truly see who you are. You chose to

Master the experience of Planet Earth, and you are well on your way to doing so. Open you heart to what it can reveal to you.

We tell you that Mount Shasta is by far the most powerful sacred site, the most omnipotent vortex-portal complex in North America. Few locations on your planet have equal capacity of multidimensional expansion alloyed in divine synergy. You are Beloved.

… and so it is. AA Metatron via James Tyberonn

Metamorphosis and Auric Maintenance 101

Archangel Metatron Channel via James Tyberonn

Greetings, I am Metatron, and you are Beloved! I come to you in this moment, a moment embellished in love, unconditional love, the love that exists as a language coded onto the very light and energies of what you term the Ascension.

Auric Fields 101

We speak this occasion on the complexity, nuances and vital importance of auric maintenance. Many who accept the existence of the Human EMF understand little about how to maintain it, the energies that impact it and the implications of its maladies. Most humans do not as yet recognize the precondition and priority relevance that the sustention and upkeep of the human electromagnetic field necessitates. Most erroneously consider it to be an auto regulated energy field that arrogates and self adjusts. The truth is that while it is capable of great self regulation and multidimensional expansion, it is only so empowered when its fundamental needs and circuitry are wholly robust and integrally supported.

The auric field, or human electromagnetic field, requires certain levels of physical, mental and emotional well being to operate in whole circuitry to insure one's growth and well- being. Dysfunction in any of these areas can create an auric bleeding effect, a short circuitry that creates varying levels of malfunction leading to an energetic downward spiral.

Interfering Fields and Bio-Polar Reversal

Now, this is indeed a timely topic. In your current times, the electromagnetics of the ascending Earth are being amplified in potent surges, pressurized surges that can stretch and micro fracture the flow and symmetry of your auric fields. This is a fundamental necessity of the Ascension as you morph into a larger EMF and achieve merkabic fields.

However, in your age of cell phones, computers, televisions and electronic devices, your offices and homes are in a constant barrage of opposing electromagnetic energy fields. The result is a temporary 'short circuit' effect that can lead to imbalance by means of energy loss through fissure cracks. In essence this may be termed 'auric bleeding.' Some of the effects are short termed, but others can become chronic.

The detrimental effects of electromagnetic fields and microwaves from your ovens, computers, mobile phones and televisions have been recognized by some of your mainstream medical professionals. These tend to be minimized and ignored by the masses, in part because the very suppliers and manufacturers of these generators of microwaves and magnetic fields sponsor and underwrite studies from scientists with the aim of disproving or minimizing the contra effects. Yet all of you are constantly within these fields. In fact most of you will read this message in front of a computer screen. Fear not Dear Ones, there are effective methods of dealing with these contra fields, and these will be offered to you.

But do not ignore the fact that these fields will have varying levels of effect on you. Most do not penetrate very deeply into the layers of your auric sheath, others do, yet all can fissure your energy fields. There may be some who consider this to be an issue of belief system or this very message to be one of fear. It is neither. It is true that an Ascended Master can transmute such effects, but Dear Ones we tell you in love that unless you are manifesting at that level and can walk on water, take heed. The mal effects of microwaves and electromagnetic waves created from alternating currents are a fact, whether you believe it or not.

While these effects are not life threatening, they do indeed affect your energy circuitry and can create what is termed 'reverse bipolarity.' We will speak more on this later within this assay.

The Crucible of Ascension Magnetics

Other less obvious but vitally important sources of auric diffusion are the very mechanisms of the Ascension. This may seem paradoxical at first glance, but indeed it is occurring, and you must be aware of this. The primary mechanisms we refer to are the solar winds, the increase in the spin of the planetary core and the resulting amplification of planetary frequencies. The planetary pulse is speeding up, and Dear Ones it is affecting you in myriad ways. In a very valid sense, what is occurring through the ascension energies is the stretching of your frequencial capacities. As such you are required to grow a new expanded auric field, much like a snake that annually outgrows its skin and goes through a phase of fitting into a new one that can better encompass the larger body. Do you understand? Before the new one can become resilient, the old one is stretched and cracks, and during that transition there is a phase of metamorphosis that contains within it certain vulnerabilities until the new one is completed.

As your Earth draws closer to the Ascension of 2012, many major changes are occurring around you that have profound effects on your physical, emotional and mental well being. Many of you find that you become somewhat 'vulnerable,' feeling overwhelmed at times, as if you cannot seem to get things done, as if there are not enough hours in the day. Some of you are going though periods of depression, you feel as though you are sinking into a funk, and it seems like you are moving through molasses. Emotions go from ecstatic highs to deep lows shadowed in dark despair. Dear Ones, you are not alone, literally millions of you are experiencing this stretching and building of the 'new auric skin,' and we say to you that by understanding the process and maintaining the integrity of your auric field, the process becomes much easier.

Case in Point

The channel has often spoken of an energetic equalization that can occur when seekers initially visit powerpoints or sacred sites. When a non resident seeker goes to a mega powerpoint such as Mount Shasta, Lake Titicaca or Glastonbury for example, many find themselves enveloped in a much stronger resonant electromagnetic field than what they are accustomed to, an energy much stronger than the resonate vibration rate of their auric field. For some but not all, within five to ten days depending on the auric strength and resilience of the individual seeker's field, the EMF will equalize in a form of osmotic equalization. The auric field will do this through the formation of tiny fissure cracks, and an energy loss will occur. Unless the individual deals with it, an auric imbalance resulting in emotional extremes can occur up until the time that the human EMF grounds, strengthens and is capable of holding the higher frequency. For people that move their residence from lower energies into the locale of a mega powersite, this process can take from six months to a year. Those of you who have moved to the Mount Shasta and Sedona areas will understand this. For visitors, it will correct itself within a week or two of going back to their place of residence. The key is to be aware of your auric frequencial quotient, and do not over-stay your time until your field is capable of holding the higher frequency.

Now, we are not saying that it is not beneficial to visit powernodes, quite the contrary. Such expansion in the overview is inevitable and quite beneficial, and it is quite necessary in order to grow the 'new skin' required to expand into greater frequency and multidimensionality. Powernodes accelerate this process and as such should be visited. These infinity points are virtual accelerators that aid your metamorphosis and prepare you for the heightened energies of the ascension, you see? Rather we are saying that there are processes occurring within powernodes that are a microcosm of the macrocosm, and that they need to be understood. You see, your entire planet is beginning to quicken, to increase in frequency, and you must ready yourselves to contain that frequency. Many of you are dealing with this, somewhat in the dark. You are now or will soon be experiencing this, and an understanding of the process will serve to make it much easier.

Auric Metamorphosis: Changing the Etheric Skin

Q & A with Metatron

Question to AA Metatron: How does one recognize auric bleeding?

AA Metatron: *First by understanding certain conditions that can create energetic imbalances and auric energetic diffusion. Those imbalances we have listed in the above discourse. The human EMF has certain built in defenses. Just as your skin has three levels of sheathing, so to speak, your auric fields have seven on the earth dimensions. The outer three levels are where most of the energetic diffusion takes place from electromagnetic waves, however microwaves can affect you on much deeper levels. Microwaved food reverses the molecular polarity of the food substance being heated. This is a recognized fact within some of your mainstream academia. When this is taken in the body,* *circuitry within the digestive system and bowels are effected, both on the physical and energetic bodies.*

Now, the precondition recognition is essential, first by knowing within what situations auric interference from opposing energy fields can occur. Sensory indications can be quite subtle. The first sensory indicators are emotional lows, a sense of being tired and out of sorts. Humans rarely connect this to EMF loss, because these symptoms occur quite commonly due to the fact that most homes contain interfering fields and due to the pressures of life within families, jobs and daily life which have their stressful aspects. These aspects indeed are in part due to energy loss from the taut hectic pace, worsened by lack of exercise and/or lack of healthy diet.

Question to AA Metatron: What can be done to strengthen and solidify the Human EMF?

AA Metatron: *There are many things that can strengthen the field. But first, try to eliminate the root source of auric short circuitry and diffusion. Stop using microwaves for food. If it is a case of being obliged to spend hours in front of a computer, as many of you do, there are steps that can be taken to neutralize the bombardment.*

Now, exercises to strengthen the auric field are numerous. Some of these are conventional methods, others are not.

1. *Exercise at least 30 minutes per day. (Tai Chi, Yoga and EMF Balancing Technique are excellent.)*
2. *Increase water consumption, and magnetize water.*
3. *Detoxify through saunas, colonic irrigation and massage therapy.*
4. *Use salt baths and mineral baths.*
5. *Use of magnetics on the soles of the feet. (A pair of*

ceramic magnets of 2,500-3,500 gauss, not to exceed 3,500 gauss, placed on the soles of the feet with one having north to the skin, the other south. Use for 10-15 minutes, no more than three times per week).

6. *Wear specific combinations of gemstones.*
7. *Wear noble metal around the neck and on both wrists.*
8. *Healthy diet.*
9. *Avoid excesses of alcohol.*
10. *Eliminate / minimize toxins such as tobacco and certain prescription drugs.*
11. *Take cleansers weekly such as raw garlic, ginger and cider vinegar.*
12. *Work with Phi cut Vogel crystals: 12, 24, 44, 88 in auric sealing.*
13. *Smudge, sage the field.*
14. *Utilize pure sonic frequency tuning forks.*
15. *Utilize the Tesla Coil Violet Ray therapy to assist energy circuitry flow and reverse blockage.*

Now, in addition to this, be aware of your emotional state. If you are suffering from lethargy, chronic fatigue, insomenia, depression or anxiety, it will certainly serve you to take the steps listed above, but other actions will likely be required. Many of you have chosen certain life lessons that involve removal of obstacles by overcoming 'contractual setups.' These setup life lessons are in essence opportunities, gifts if you will, to allow you to move forward. If they were easy, you would not necessarily learn. Simply changing your outlook to 'positive thinking' when you are dealing with a life lesson that leaves you in a state of lethargic depression is not enough. A 'Pollyanna' outlook will not resolve the core issue, none the less an intact auric field will assist in overcoming these issues, and in certain cases the auric field will not be whole until these chronic issues are resolved. One provokes the other and vice versa. (Mickel Therapy is a recommended approach for assisting with such setups that result in energetic disorders.)

While some of you may be somewhat 'karma free,' most of you still have learnings to walk through and issues to clear. The current time is a gestalt to allow you to achieve these lessons and to confront and remove remaining energy-viral obstructions. Auric maintenance underpins all.

Question to AA Metatron: Can you advise if the 'neutralizing chips' that are available to eliminate detrimental effects from monitors and computers are functionally valid?

AA Metatron: *They are of assistance, but that technology is not yet of a sufficient advancement to be very effective. The most beneficial means to deflect these fields is tri fold, the first point being applicable in all conditions of field strengthening to negate interference:*

1. *The wearing of gemstones is far more helpful than the current neutralizing chips. A single refractive gem of 2 carats or more worn on one hand and a double refractive gem on the other helps deflect the fields. Examples of single refractive gems are diamond, garnet and spinel. Double refractive gems are aquamarine, sapphire, ruby, emerald, tourmaline, topaz or any of the quartz varieties such as amethyst, peridot and citrine. In reference to the double refractive, a carat size of 4 or greater is better. Combine this with noble metals on each wrist: gold, palladium or platinum is best. Silver, copper, brass, titanium and carbon steel will assist if the more noble metals are not affordable. Wear a chain around the neck with a stabilizing pendant, such as lapis lazuli, malachite or azurite. Through this process you have a greater ability to increase your field and deflect opposing ones, you see.*

Clear gems are great producers of higher dimensional light waves, amplify one's field and help hold it intact. Do not wear beryl (emerald, morganite, alexandrite, aquamarine) and corundum (ruby and sapphire) at the same time.

2. *Try to maintain a distance from the screens. For computer monitors this is difficult, for televisions, four to five meters is recommended.*
3. *Placement of ionic generators, such halite salt blocks and air filters are beneficial in restoring the anion to cation ratio in rooms that contain computers, microwaves and televisions.*

Question to AA Metatron: Some metaphysicians and shamanic teachings speak of 'psychic attacks' occurring when the auric field is 'open.' Can and does this occur?

AA Metatron: *It can and does indeed. However, your term of psychic attack is in truth an aspect of electromagnetics and harmonic oscillation. From a higher perspective these are part of the setup and overlay of growth in duality as you learn to be responsible with your innate creative forces. As we have told you, your thoughts and emotions have a vibratory frequency within duality that is quite real and quite alive. When you focus on an event or react in strong emotion to another person or situation, you create energies termed thought forms. These amass harmonically in various manners. If you project great charges of emotionally charged energy, such as anger, jealousy or love, passion and joy toward another, you will agree that both parties are affected, yes? If you consciously dwell on these charged vibratory thoughts, then the energy mass, normally short lived, can gain sufficient energy to become an aware thought form. When spiritual love and compassion are projected and created, a synergy occurs that benefits both parties. When fury* *and hateful negativity are projected, a toxic reaction can occur from both sides. If the auric field of the 'targeted' person is open, the effect is worsened somewhat, and a temporary energy bleed can result.*

Now, it is important to differentiate between malicious or controlling energies and honest reactions. If someone disrespects you, the appropriate honest reaction can be one of pure anger and hurt. When these are openly and honestly expressed, a cleansing gestalt occurs that can lead to better understandings and a renewed improved communication. It is a learning process and is appropriate. But none the less, a form of energetic opposition takes place.

However, when one person feels a vendetta, or a prolonged desire to control the other, the gestalt can dissolve into charged conflict, and as such a malicious energy battle of will may occur. Very often both parties feel 'right,' and the frequency of hate harmonically attracts more and more like energy until the amassed energy form is so potent that it has the ability to effect a destructive 'attack' role to both. Unless one or both of them sees the wisdom of releasing the hate, it will pull them into a downward spiral, creating dis-ease and deeper negativity.

When one is in a state of malicious anger, self loathing or depression, that vibratory resonance on its own will deplete and open the auric field. It becomes a self dug pit that gets deeper and deeper. Lifetimes can be wasted. Yet on a higher perspective, much learning occurs. There are times when great souls choose life lessons of overcoming such energies. Some evolved souls such as Gandhi and Nelson Mandela chose setups to be wrongly accused and convicted of crimes in order to spend years in the negative energy of your prisons to learn to find peace and strength of will under the harshest conditions of oppression.

Now, when you are in conflict, insure that your auric field is maintained, and do not allow yourself to fall into malicious reaction. It is not wrong to react in honesty, but the narrow path of mastery is to not fall into hate and malicious revenge. You see, getting caught in these traps is much easier than getting out of them. Hate attracts more hate. When groups oppose one another, the collective of like thought forms amass, blend and oppose one another in enormous conflict. Very often entire warring nations form collective energy fields and reincarnate in these groups to continue these conflicts until they are finally resolved. Your current Gulf Wars are such an example, a roll over of the Crusades. Your World War II was a continuation of the Atlantean conflicts between those of Poseida (Law of One), and Aryan, (Sons of Belial). Very often the most difficult 'attacks' are due to one's own negative thought forms coming home to roost.

Perhaps the most difficult of these is around learning self love. When one falls into depression, self rejection or self loathing, the attack is self generating. The thought forms these dear souls create can become so potent that the auric field splits into personality fragments. A figure eight pattern of negative energy is emitted, amassed and flows back through the opening of the solar plexus. The thought form of self hatred achieves a level of awareness and will become a very real obstacle, a self imposed dungeon, until the person learns to face the root of the problem through great effort of will and wisdom.

Energy exchanges occur quite often in daily life. Some are mutually beneficial, others are not. It is important to note that energy cannot be taken from ones whose aura is intact. Healers constantly give energy within <u>an intact aura</u> and are essentially unaffected by the transferal. As such, love energy is passed from higher sources, and the healer's auric field is capable of *willingly being the conduit without losing any of their auric field energy.*

Now, anytime you feel anger at someone, those thoughts are projected. Anytime someone tries to control you or vice versa, an energy projection is launched. There are of course situations in which it is appropriate to follow another's direction. You do this in every aspect of your life. It occurs in offices, militaries, schools and between children and parents. This is based upon agreement and is properly germane when not abused. However, it is not befitting in these and other circumstances for one to allow another to abusively dominate their spirit or maliciously attempt to break their will. Abuse of power often occurs in relationships, marriages, in work, family and social scenarios. In certain scenarios as described, it is wrong to allow another to abusively impose their will and is equally wrong for one to seek such control outside the appropriate structures of agreement. This can evolve into a form of, in your terms, 'psychic attack.' The controller attaches to the solar plexus center of the controlled party and literally takes their energy and interjects a destructive domination. This 'vampiring effect,' in your vernacular, especially occurs among egocentric people, controllers and manipulators. It often is attempted unconsciously from people in imbalance and depression who need a 'lift' from being around others due to their own energy shortage. It also takes place on a larger scale in patriarchal dogmatic religions, male dominated societies and in marriages. It is more difficult to counter in such trappings. When one is aware of being the target of malicious energy projections, auric integrity is of vital importance. The visualization of enveloping 'white light' is the generally accepted protective mechanism. However if the auric field is open, the light visualization is not enough. The procedures listed above to strengthen and seal the field should be utilized. Be aware that anytime you have strong emotions of a negative nature or dwell in depression, your fields will temporarily fissure. These attacks can only be energetically depletive if the field is open. The projected negative energy is easily repelled when the auric field is wholly intact. In such 'wholeness' the energy is reflected back to its source for the sender to deal with. There is lesson here, and Dear Ones, do not be the sender of malicious energy, inevitably it will come back and cause you great remorse. Such is the nature of the law of harmonic oscillation.

Religions are often the source of great inappropriate control, control through fear. Even with the 'New Age,' in your terms, have sprung up gurus and spiritual teachers whose fame and power leads to the downward spiral of ego, self-aggrandizement and control. The path of leadership and power inevitably forks, and one may be blindly tempted to take the path of greed and power over love. It is part of lesson, and many have fallen into such traps of ego. When this occurs, they become 'energy takers.' That is why you should never blindly follow any leader or channel. Rather use discernment, and attune to your own Divinity. When you become a part of any 'group consciousness' and then decide to break free, there is a natural pull from the collective to bring you back in, and as such a form of 'energy attack,' in your vernacular, occurs, especially in possessive collectives who work at recruitment of followers.

Question to AA Metatron: What is the difference between an auric field (EMF) and merkabah?

AA Metatron: *In your present times, many of you think them to be one and the same. They are not. Both the aura and merkabah are electromagnetic fields, but different stages of same. The merkabah is the crystalline, geometric form of the auric field. Fissure cracking and energy diffusion cannot occur in the merkabic phase. However, few of you retain the geometric* *crystalline merkabah form 24/7, and a fissured bombarded aura cannot easily transform into the pure merkabic form. As you learn to maintain your field, you will find it easier to achieve merkabic states, the initial phase of which is the star tetrahedron. Merkabahs evolve into even higher geometric forms beyond the 8 pointed star.*

Question to AA Metatron: You mentioned combinations of specific gems for strengthening the auric field. Can you elaborate on this?

AA Metatron: *This topic is a vast one and a book on its own. Briefly, gemstones, relative to auric maintenance and strengthening, are in essence benevolent conscious generators of force fields that reinforce and fortify one's own EMF. Being crystalline in matrix, they also are tuning forks that assist in merkabic formation and expansion. A basic combination would be to wear a single refractive stone, such as diamond, garnet or spinel on one hand and a double refractive on the other. The best single refractive is the diamond, but it needs optimally to be a solitary of at least 2 carats, and we realize these are price restrictive. The best substitutes are garnets, preferably the Ural Mountain green demantoid, or a red-orange spessartite. Both project the octahedron in crystalline form. The diamond projects the dodecahedron and octahedron. If one can afford it, a combination of colors among single refractive is best. These can be intuitively alternated according to astrological forces and one's cycles. Garnets come in virtually all colors except blue. The spinel comes in blue and violet.*

In terms of double refractive gems for rings, the most potent are ruby, sapphire, emerald and aquamarine, projecting the largest fields assuming a 3-5 carat size. Alternate colors as cycles shift. Tourmaline, topaz and quartz varieties such as opal, amethyst and citrine are all piezoelectric and also quite potent force generators. Gem use is not folklore, indeed they are crystalline force field generators of astral light. They can increase one's vitality and even prolong life span, particularly when used in tandem. Well to study this topic. Remember the body is bisymmetrical, hemispheric. Combine gems with metals about both wrists and the neck. Noble metals such as gold and platinum are most potent. Silver is tertiary but quite benevolent in its aspect. Alternate these. A pendant about the neck is also recommended. Study this topic, there are numerous books available.

Closing

Dear Ones, I leave you with this message, trust that all is well. The changes coming through the mechanisms of the Ascension are for your benefit. You are changing your etheric skin in order to encompass these magnificent expansive qualities.

I am Metatron, and you are Beloved!

…and so it is. AA Metatron via James Tyberonn

The Flashing Universe

Archangel Metatron Channel via James Tyberonn

Greetings, Beloved! I am Archangel Metatron, Emissary of Light! I welcome you here and honor each and every one of you who read these words within your own unique vector of NOW! Indeed you are living in a time of special awakening, a time of special rediscovery! You find yourselves in a unique sojourn within space-time sequence, one that will allow for quantum leaps in your consciousness! You see, your Ascension is truly an event of multi-dimensionality, an event that may be described as holographic.

Now we will speak on multidimensionality, time and the mechanisms of the flashing universe. Lifetimes, in your terms, are not linear, rather they coexist in the Unified Quantum Field that we term the eternal NOW. In truth everything you experience, including the very moments in your lifetime, are not linear. That is an absolute truth that most of you have difficulty imagining. So, you may ask how can that be? I am aware of what happened yesterday and last year, and the results of many things I planned came into effect because of the linear sequence, the build of my past actions leading to my present.

The Metatronic Field

Dear Ones, we tell you that the nature of your Universe and all its myriad of probabilities coexist both above and within the waves and particles of consciousness. The Universe as you know it flashes in and out in a pulsing sequence of light energy frames that consist of what we term consciousness units. Your universe, indeed the Cosmos, is in a constant flux, an infinite flash between matter and antimatter. That Divine Flash is at the very basis of the Metatronic field of 'All That Is.'

In your present world, in your level of the NOW, you are only beginning to rediscover, explore and grasp the 'flashing' matrix of creation of what consciousness is, of how your mental processes form reality through waves of conscious energy and indeed how to manipulate this for greater growth. You constantly are doing this on what is termed the unconscious divinity level and consciously by learning how to transform your thoughts and your creative emotions into matter, into physical reality. Most of you perceive and measure your inner development by what you see outside of yourself in the exterior events and horizons that make up your exterior world. But the reality is that what you see on the outside is a direct result of what you think and believe, of what you mentally create.

As humans you tend to think of your greatest component as your soul. The soul composes the life of each of you, in these terms, but we tell you Dear Ones that you are more than your soul. The soul evolves and thus changes with each pulse of your creative beliefs and thoughts. The soul is an instrument of your growth. The soul is in Metatronic terms an attribute, a vehicle of your growth into Divinity, of God source.

In a sense, it is the merkabah of your divine self, the oversoul. Accordingly your soul is constantly changing its structure as you grow, yet retaining the identity and integrity of the expansive individual as it connects more firmly, more consciously to that which you term the oversoul. You see, you are constantly protracting the soul in flashing awareness, and in so doing you increase your capabilities within and beyond your accepted realities. As such, you expand your system of belief and thus expand your creative grasp of reality. You see, there are as many realities, as many universes as there are systems of belief.

Reality then, occurs on a fertile matrix. The process of God Source seeding the matrix of awareness through pixels of minute particles involves what we term flashing units of consciousness, electromagnetic units that can and do pulse in and out of all realities in a strobe light fashion. These flashing energy units are geometric in nature and occur with a relatively short 'life' span. This is a Metatronic process above the level of the form in which I address you in this NOW as Archangel Metatron. It occurs on the level of Lord Metatron far above the level of the 12th dimensional field. In a manner of speakin, I as Archangel Metatron, am to Lord Metatron as you are to the Archangels, far beyond the dimension of 12 within the realm of the Creator God. In this realm, Lord Metatron exists as the conscious generator of all pulsing creation, flashing within and without sacred geometric energy units that can be termed as consciousness codes in the form of geometric light particles into all universes and parallels in time sequence probabilities.

Black Holes and White Holes

That which you term Black Holes and White Holes are the mechanisms of this process. You see, the field of your reality, of your universes are in a real sense folded within themselves somewhat like the analogy of a two headed snake. One head consumes while the other regurgitates that which has been digested and consumed. This occurs in an infinite pulsing process between matter and antimatter, and we will explain later in this discussion.

Now, your scientists postulate that black holes are imploded supernovas of extremely compressed energy whose gravitational field is so immense that it pulls into itself all matter and light. Yet they do not as yet recognize that at the back of the black hole, in a manner of speaking, is a white hole that generates new vitalized energy into the universes.

Although your academics accept the theoretical existence of black holes, they do not agree with the concept of white holes. In fact both exist and are intimately paired. Both in fact exist in the universe in macro and micro.

At the centre of your Cosmos is a great central complex, in Metatronic terms, referred to as TULA. It is a great centralized manifold of a macro black-white hole circuitry. It is the mechanism of SOURCE. It emits the frequency of the Universes in a geometric fabric. It imprints gamma grade light, encoded with an essential resonant frequency known to you as LOVE.

Indeed your sun and some stars are white holes, feeding streams of charged energy from nuclear fusion through solar winds into your solar system. However, most of your scientists conclude that white holes are outside the laws of thermodynamics, and that they refute the mathematics behind the physics. So they stand 'safely' behind the shield of such limited conventional thought. Yet, in past decades they have seen numerous 'perfect' mathematical theories crumble into astrophysical dust. You see, the conundrum is that mathematics will never work in astrophysics until you switch from base 10 to base 12. All in time. Indeed, it was not so long ago that zero was even considered a valid number by your mathematicians!

Your Einstein theorized the existence of black holes, but it was perceived as an imaginative unicorn. Dear Ones, these unicorns exist! Could it not be that if you consider a black hole to be a collapsed star, then a Sun or Super Nova may be a white hole? Your sun is much closer to a white hole than academia currently considers. Yet it is easier for them to digest than the 'thermodynamically impossible' matter-energy portals you term wormholes.

You see, black holes and white holes are connected in a manner of speaking by a wormhole like conduit, a spiraled tunnel that connects them through parallel dimensions. Black holes and white holes are the two headed snake. You cannot have one without the other. White holes are in reverse spin at the backside of black holes and vice versa. Black holes are the vacuum. White holes regurgitate renewed quanta of energy in the cosmos through massive nuclear fusion. Both are multidimensional. Indeed within other universes there are black holes drawing energy, and at the opposite end of it is a white hole firing energy into your universe. You see, light and matter in the other universes enter the black hole and are disgorged outward through the white aperture into your universes through supernovas, stars and suns.

Do you not see the logic of this? Every system, from physical organisms to galaxies and universes, has its intake of energy units and output of energy units. There are organs and systems within your physical body that intake and emit energy. Your planet has intake points and projection points of energy within vortex and portal systems. This occurs in everything, both on the micro and macro level. In a manner of speaking, even the cells in your body contain black and white hole energy systems.

Electrons as White Holes

Now in the micro, some metaphysicians have suggested that electrons are black holes and protons are white holes. This is on track but not exact; it is in fact the reverse. Electrons correlate to white holes and protons to the black. Electrons and protons are the receivers and regulators of the flashing universe sequence created between the in and out pulse that occurs between matter and antimatter through black and white holes. Black and white holes are the mechanisms of the flashing aspect of the Cosmos. One flashes energy in, the other flushes it out in a rhythmic cosmic respiration. The out flash sends a massive surge of conscious light energy into universes, and all living things perceive it through what may be termed an attention unit or an awareness flash. Immediately after the out flash there is a reverse inward flush where energy is taken out, and during that phase consciousness pauses; it blacks out for an infinitesimal unit. Then the cycle repeats itself. Electrons are the receivers of the incoming white flash, and protons are the regulators of the dark instant out pull counter flash. Electrons are the instrument of white holes, protons of the black holes. Yet truly both are tied at the hip; they work in amazing synchronized tandem.

When the electron receives the consciousness flash, the proton on the other side receives and triggers the void moment that occurs between out flashes, both spinning in orbits that thread into other dimensions. The pulse fueling all of this is the cosmic heartbeat that oscillates between matter and antimatter.

If you want pure symmetry, then a black hole would have negative mass whilst a white hole would have positive mass. Now, think what you would get if they were threaded – a zero mass unit connection of electrons and protons effectively pursuing one another, pulsing, strobing in and out of your world in opposing spin! As such, there is a constant surge of new energy into your physical and nonphysical reality which comes through a nearly infinitesimal number of black and white holes. The sequencing of energy in and energy out creates a strobe like flashing effect within the consciousness and time sequence of your reality. Your oversoul is calibrated within the flashing sequence.

Imagine then, that the root unit of your beingness, in your terms your oversoul, is at the center of a wheel. This hub is a flashing receiver of conscious light and emits an infinite number of pulsing light threads or 'fiber optic spokes,' and at the end of each spoke is one of your soul aspects or incarnational sojourns. The reincarnation aspects represented by the spokes of the wheel receive units of conscious light termed consciousness units that are flashed outwardly at an amazingly rapid strobe. Each pulse simultaneously garnered in the oversoul is individually sequenced to each lifetime. These pulses make up the awareness unit within each life. You are only enabled to be individually conscious when the attention unit is pulsed. These seem to be sequenced and linear, but in fact there is a great timeless void in between each pulse.

Now, as each spoke or lifetime aspect becomes more resonant, more capable of holding the energy of energy pulses, it consciously expands back into the center and becomes aware of the unity of the multi aspects of the hub and all the other spokes, and in so doing it evolves back into the center, becoming part of the center. But all is achieved in a matter-antimatter pulsing sequence, a flashing strobe aspect.

The soul aspect becomes aware that it is coexisting on many layers of time sequence, and that they coexist both within and above the pulsing aspect of time sequence. On this oversoul level, you are in the conscious eternal NOW.

Forward and Backward in Time

Therefore achieving the ability to be fully conscious in the NOW, one can literally go backwards and forwards in time and can be in all places and in all times simultaneously. The soul aspects can combine to create complex gestalts of consciousness, and their unpredictability allows them to form infinitely varying patterns, potentials, probabilities and fulfillments. Do you see?

A different scenario then presents itself, and further growth and challenges occur. Yet even as your consciousness graduates into attentive awareness of your greater identity as part of the Co-creator God source, there always remains another level above to grow into, another part of the never ending symphony to play. You see, beyond your oversoul, beyond your divine Self, as you think of it, there is another higher self and beyond that still another of which your oversoul is dreamingly aware. And that self tells you that there is a reality beyond human reality and experience that cannot be made verbal or translated into human terms. Truly all reality, all consciousness has no beginning and no end. I seed you herewith the desire, in your terms, of grasping the understanding of this valid concept. You see, there is indeed an expanding universe, and it is ever formed, ever created in the eternal NOW of your present. Simply stated you cocreate your own reality, and the perception of that reality is multidimensional and beyond what you are capable of perceiving from this stance. Yet I assure you that it is the one truth, and it functions in a pulsing strobe. It is created in a pulse shift that works and frankly speaking is true whether or not you care to put your mind around the concept of these expanding realms.

Now, the mechanism of this form of consciousness involves crystallization awareness, geometric patterning of the thought process by means of what you term impeccability, Purity of thought, thought that is in image unit, particle and not wave.

Your existence on the planet of duality upon the dimensional planes of Earth are somewhat limited in perceptional lenses, somewhat filtered by design. An analogy of the human brain functions can be seen as akin to a dual 'radio' receptor. One aspect, the ego-personality, can be considered to be in the band of AM radio, and the subconscious or God Mind aspect as tuned into a broader FM width. This indeed has its own purpose, and because of these preset limitations the personality-ego perceives and focuses into reality a single event from all of the probable ones. Until the ego experiences the event, it is only one of all the other probable ones. It becomes actual in your reality only when the physical aspect of self experiences it.

A portion of the brain, the subconscious or greater SELF, can and does experience reality in an entirely different fashion than the ego-personality brain does. From that perspective, from that integral higher stance the subconscious is fully aware of all other dimensions, probabilities, potential events and time sequenced realities separate from the one experienced by the frontal brain aspect of the personality physical base or ego self. And in following the AM/FM analogy both receptors, both perceptions are equally valid, but one is tuned to only one station and with reason. That reason is what is termed focal life lesson and indeed is part of your growth process but one in which you are destined to grow beyond. In truth, you are like Dorothy in the metaphoric tale of "The Wizard of Oz." Learning to access the infinity of the higher mind comes only after learning to willfully create with the ego mind. Indeed when you do, and you already have in a real sense, you will realize that you had the power all the time. You simply needed to 'click the shoes together,' as did Dorothy of Kansas in her path of exploring the wonders of Oz along the yellow brick road. Do you see?

Now, the greater part of you experiences multidimensionality within all potentials and parallels with an equal amount of robust exertion as does your physical self in the experiencing of narrowed focal physical reality. The myriad of all the other parallels and probabilities are indeed quite as experientially beneficial for your 'soul growth' and indeed occur with commensurate actuality in not only other dimensions of parallel reality but simultaneously in what you term other lifetimes. There are an infinite number of systems and universes between potential and physically actual systems of matter and what we term Metatronically speaking as antimatter. Using only the limited physical senses you can never perceive these other systems, yet probable events and parallels are substantive bonafide experiences even if you do not perceive them with your physical antennae. In these other systems of reality, each of you has 'other selves,' not as an identical self but through different aspectual parts of your oversoul. These are developing abilities in a different way than you are in physical reality, emphatically contributing to your soul expansion in consciousness. Do you understand?

Now, you indeed often experience probable events and parallel worlds in the dream state in a conscious manner. These experiences cannot be readily or fluently translated from the greater mind into the narrow AM band of the conscious ego mind. Yet they can be translated and brought into wider awareness in the frontal mind through concerted disciplined effort in development of the inner senses. There are schools of disciplines that have taught these techniques through various modalities throughout your history in what is termed 'Shamanic Journey' and out of body experience in meditation and in various orders of schools of wisdom. These systems of journey into probabilities are quite as real and quite as educational as what you term your physical system; you are simply not focused within or motivated to develop them.

We have said that there are as many universes as you care to imagine. To expand on that, and please forgive the pun, it is the nature of your dilating group matrix to create as many universes and parallels as suit you. Indeed there is no one format, no one template. Consciousness is source, and consciousness of which you are a cocreating participant indeed creates space and equally dissolves it within the validity of the collective imagination process of creativity. Universes are a constant tapestry that you eternally weave in conscious creation. There are no limitations, no borders and no set order. In fact the 'All That Is' in your now is quite different from the 'All That Is' that you created yesterday and from the one you will create tomorrow. Consciousness is by nature creative, and it has no beginning and no end, in your terms.

Space as Congealed Holograms

Space and matter are seen as congealed thought and time as sequencing of same, appearing to flow, appearing to be linear but in truth neither. Imagine time as a movie made up of multiple individual still shot frames, moving through a focal light lens. Each is truly separate, but the process of seeming linear sequence allows for a unique focus that enables singular understanding of the frame-worked progression. The observer has the ability at any time to edit, play back or fast forward to any part of the film, to change it, relive it or discard it. Likewise, units of consciousness, pulses of consciousness are truly outside of time, they are the observer. They do not occupy space-time. They are truly not of your dimensional world, not subject to the constraints of the 3-Dimensional lens of reality and are not accurately defined by your current theories of relativity and time-space continuum. Consciousness and attention units of consciousness can move forward or backward in time line harmonics and do so at speeds far greater than the velocities of light.

So it is perhaps hard to grasp, but in between the moment when you first began reading this channel and the time where you are now, your consciousness has pulsed in and out into living attentive frames of all of your coexisting lifetimes including your own future and past, in your terms, within this one. Your own physical and etherical chakric systems contain within them miniscule white holes and black holes that allow you to do this. The ongoing insertion and discharge of consciousness energy units is indeed the very process through which you experience the purposeful 'illusion' of consecutive time.

Even that which you perceive as your planetary history has an infinity of variables, all of which are valid. Your lifetimes move forward and backward in time. In your terms, you may have a 5th lifetime in time frame millennia ahead of your next. That is an example of how the holographic aspect of time is experienced. The channel speaks of having several friends all of whom may relate to the period of King Arthur and Camelot as having been King Arthur or Sir Lancelot. Or to site another example, knowing four or five people all of whom felt they were the Disciple Peter at the time of Christ. The fact is that it is quite possible that all of them were Arthur, all of them were Lancelot and all of them were indeed the Disciple Peter. How? By means of the timeless holographic drama inserts that are available to be lived and relived, forward and backward in time, by as many who wish to have that learning experience in the living hologram. Do you see?

We will tell you that aspects of the Golden Age of Atlantis are in fact from your future, in your terms a future that in this frame has been seemingly pulled into your past! Is this not fascinating?

Now, even the physical objects that surround you within your chosen version of the NOW are collectively created by you and can be seen as flashing symbols that accurately express your subjective level of growth through the current thoughts and feelings. The same is true of the particular version of world events that you presently pull into the holographic experience, whether war or peace, prosperity or abundance. It is but one of many potentials and reveals more about your collective state of group consciousness that you may wish to admit.

When you learn to tap into the flashing pulse of creation units through your 'higher beingness,' you will be able to travel consciously within and without the width and breadth of space and time and all of its parallels. To do this, you must understand that this will not be achieved in your normal ego-personality state. An AM radio will not receive FM waves. You must learn how to go into the Quantum Field of Unified Consciousness, an aspect that is within all of you, and learn how to consciously navigate within it. All of you are doing this constantly from a higher aspect, but to be able to navigate with conscious memory in what may be called lucid states requires focal development of the unconscious field. To even begin the initial workings toward the attempt of this complex feat, one must first convert the auric field into the geometric merkabah. Do you understand?

Now, visualize yourself as part of the infinite universe, one in which all of the planets and stars are conscious and teeming with indescribable energy. Think of this universe as having the form of a body. Imagine its outline brilliant against the sky. The suns and planets are your cells, each filled with energy and power but awaiting your direction. Then see this image exploding into your own being, into your own consciousness in unbelievable brightness. Energy is in this way directed and is constantly received and emitted into the body in flashing units. Realize that your true SELF is a portion of far greater multidimensional structure, spread out in even richer dimension.

In your current carbon based biophysical existence, you are learning how to handle the inexhaustible energy that is available to you, and indeed there is much to assimilate. Like Dorothy of Oz, there are tools available to you that you do not yet recognize or utilize. It is time to activate your crystalline nature, Dear Ones. One of these potent keys is the utilization of mineral crystalline forms for attunement of the energy field into its geometric merkabic crystalline form. Indeed crystalline forms are more capable of holding the pure sequence and energy of these units and indeed of assisting in the axiomatic merging of the front mind into the wider realm and frequencial sequencing of the God Mind harmonic. Gemstones and phi cut quartz such as the Vogel crystals are helpful assistants in this process. Indeed, crystalline forms geometric in molecular pattern can aid in tuning the human energy fields more harmonically between dimensional fields as well as through the void spaces of consciousness flash sequence. Gold and platinum also are tools that benefit this process.

Consciousness and its flashing nature and units truly do not have human limitations, but they indeed offer holographic opportunities for growth, experience, drama, learning propensities and lenses of all probabilities and possibilities. They are vitalized and validated with all the divine qualities of being sacred fragments and perceptions of 'All That Is,' 'All That Was,' 'All That Might Have Been' and 'All that May Be.'

I am Metatron, and I share with you these Truths. You are Beloved.

… and so it is. AA Metatron via James Tyberonn

The Inner Earth and Her Kingdoms

Archangel Metatron Channel via James Tyberonn

Inner Earth

Greetings, Beloved! I am Metatron Lord of Light, and I welcome each of you to these living words. I embrace each of you in love.

Now, you think of your planet as being a solid sphere, but, Dear Ones, indeed your planet is flattened at its poles and contains great chasms within it that house what you term lost civilizations. Your planet has been determined by your geologists to be 4.5 billion years old, and we tell you that is close to being accurate within the laws of your science. What your scientists do not understand is that dimensionality influences the laws of your science. From a different dimensionality your planet would be a conglomerate of many Earths, concentrically juxtaposed with an infinity of parallel probabilities, you see.

Question to Metatron: Much has been said about underground civilizations. Do these exist and if so where?

AA Metatron: *Indeed they do exist and have for a very long time. There is a race of humanoid beings that originated at the time of Lemuria, and at a later phase in Atlantis, who found means to enter into the hollow chasms of your Earth. There are huge hollow expanses within your planet that were once connected by vast tunnel networks. Those beings that first entered were of a lighter body form, a physical form not as dense as that which humans have since evolved into on the surface. Those of Lemuria initially entered into these crevices to escape the rampage of what you term dinosaurs that overran their continent for a long period of* *time. This was how they learned that an inner sun, in a manner of speaking, exists as they ventured deeper. The inner sun emits a form of blue light, and the ancient Lemurians in semi physical etheric bodies developed a means to see within this light, discovering an amazing beauty in the chasms. Some of you think that the Lemurians ascended. The truth is that they descended into the Earth, but they are very close to what may be termed their completion, or Ascension.*

Question to Metatron: Where do these beings live, and what are they like?

AA Metatron: *Indeed, these beings of Lemuria are far more highly evolved in spirit than you are at this time. Their bodies are, as I have said, less dense but are indeed physical. Their skin has taken on the hue of green and in some cases a green blue, because the water they drink is highly mineral in content, containing oxidized copper and other metals in a higher concentration. These beings are peaceful, knowing no religion other than the love of source and an understanding of what is termed great tranquility. They are quite aware of you but have no desire to intermingle. Why, you may ask? Because of several reasons, the most prevalent is that they are aware of your violent nature, your fear and your lack of mass spiritual development. They are aware of your physical diseases, some of which could potentially infect them, and they are quite unable to sustain your solar rays. Let us say they have evolved into beings that have set a unique course and are close to completing its sojourn. Their bodies are sustained by a crystalline magnetic force and by a light source that is emitted from your planetary core. Their mental capacities have allowed them to harness these force fields and adapt them to their physical and spiritual sustenance. Although they are in physical biology, it is a 4th dimensional biology and thus far less dense than your own. Many of them are quite adept in what you term teleportation.*

There has indeed been contact between the inhabitants of the inner world and your governments but not an intermingling. There have been messages of the need for the human race on the surface to be aware of the coming changes. As we have said, those of the inner world are on a different course, a different time cycle and one that is close to completion.

You have asked where these inner chasms are located. Keeping in mind the dimensional parallel aspect, we answer you that in your terms the greater concentrations of these chasms are in lands below the seas, but there are pockets that exist below every landmass. The northwestern regions of the United States, the mountains of Arkansas, New Mexico, Arizona, Mexico, Central America, Peru, Britain, Europe, the Himalayas, Chile, Argentina, Bolivia, Brazil, Central America, China, Siberia, Greenland, Iceland and Sri Lanka all contain subsurface colonies.

Question to Metatron: How deep do they live within the Earth, and does the Earth have access points to these inner hollows?

AA Metatron: *We have stated that entry points occur at the poles and in many apertures upon the planet's surface. It is impossible to define how deep, as indeed they are parallel, and measurement in your physical terms is not truly valid. But suffice it to say that the great hollow chasms in your terms would be some 20 - 30 miles below the mantle surface and as deep as several hundred miles. (Your geologists would find this hard to imagine because of their interpretation of pressure and heat, but we tell you these places are real, and life exists in them through regimenting force fields.)*

But these distances are relative to multidimensionality. If, for example, you were to enter one of the polar dimensional regions or one of your triangulated dimensional passages into a parallel Earth in a different space-time continuum, how far would you truly be from the point you entered it? The answer is in inches and eons, depending upon the stance from which you view it. The closest science to accurately encompass the infinity of worlds is quantum physics, and that science is only beginning to be understood in scientific communities on Earth. The paradigm must change as well as the geometric and mathematical base before this science can be accurately deciphered on your plane.

The primary entrances to the inner hollows of the planet are through the Polar Regions. The expanded dimensional aspect indicates that your multidimensional Earth(s) are quantumly connected at the poles. It is the magnetically flattened aspect of the Polar Regions that allows for the hollow chasms in the planetary sphere. Your Admiral Byrd wrote of flying over this region and seeing tropical lands of flowing rivers. We tell you that he did dimensionally pass for a brief timeless period into the parallel aspect of the hollow Earth and that he and others did so on several occasions. At the Polar Regions, dimensional hyper transport can occur under a range of specific conditions. The polar aperture dimensional transport occurs somewhat like traversing a steep conical inclined field and elliptically flipping. The sensation is that of suddenly being 'flat' again, or parallel to the surface. In fact, one is then indeed in a parallel dimension. Those in the inner Earth exist in such space. Many races and beings exist in your 'multidimensional' Earth, including many that you term extraterrestrials who would claim citizenry of your planet just as convincingly as you do. Indeed they have in most cases been on the Earth longer than mankind.

Question to Metatron: Does this include the beings we refer to as Yeti and Devic?

AA Metatron: *It does not. These beings also exist but are not among the Lemurians and early Atlanteans that ventured into the inner Earth.*

Question to Metatron: Can you tell us about the Yeti, and are these the same as the beings called Bigfoot?

AA Metatron: *These you refer to as Yeti or those termed Bigfoot are earlier versions of the genetic experiment on Earth, primarily from the phase of Atlantis some 200,000 years ago. At that time there were many genetic experiments on your planet. These beings are intelligent but genetically impaired. These massive forms were genetically created using human DNA with that of the ape to create a laboring humanoid beast, a beast with greater intelligence, human like intelligence but with imposed genetic wiring or implants that 'unplugged' certain areas of the brain. These are survivors of what you term 'the others' of Atlantis. These are remnants of beings cruelly created for work force within mines, farms and forestry. The areas of their brains that allow for emotion and expansive thinking have been artificially impaired within their codes. Yet it is the same source inhabiting these diminishing creatures that is within the dolphin, although within the Yeti bodies it is unable to find expression or mentally evolve beyond certain limits. Yet they do possess that which is termed divine intelligence. Only within physical strength and survival instinct mechanisms are they able to sustain and grow.*

These beings do not live in the inner chasms of the Earth. They live in caves, remote mountains and within your deep forests and swamplands. They are nocturnal beings. They are a race that is dwindling, shall we say, and in time will no longer exist. These beings have great fear of humans and experience a profound sadness and confusion as to their evolvement. They cautiously watch you with baited anxiety, they know they are your brothers and want to be closer to you but intelligent enough to know they cannot. Their bodies have evolved to allow them through the massing of thick hair and thick oily skins to live in extreme locations, and here they live out their final era of life. Their spirits no longer wish to complete, as their genetic limitations are such that they cannot evolve. If you were to look within the eyes of one of these beings, you would feel a great sadness.

Question to Metatron: Are you saying that Yeti are the same spirit source as dolphins?

AA Metatron: *Indeed. But be clear, it is same source not the same expression, in fact far from it. You see, these creatures do not express the same vivid emotion, intelligence and joy as do the dolphin, because the bodies in which they are entrapped do not allow them to. You may wonder why such spirits would choose to inhabit such restricted physical vehicles, and the answer is that fewer and fewer choose to do so, and soon they will no longer exist as a species. These beings are capable of expressing great strength in survival and love for one another but have decided to vacate this expression. If man continues to soil the oceans and slaughter the magnicient energies of whale and dolphin, so will these magnificent brothers vacate this role. You have no idea of the light energy that whales and dolphins exert into your seas.*

Would it surprise you to know that many of you exist as dolphins, both in the present and past? You do so to perform a great service to the planet and to mankind. You see, the grid alignments, powernodes and white holes that exist on your planet do not just exist on dry land; most of them exist in fact over and under waters. Dolphins primarily and in a few cases whales align their energies to these sites in assisting to balance the energy of your planet. That which is termed the Devic Kingdom has far more life expressions within the oceans and lakes than on rock, land, plant and soil.

Question to Metatron: What is the Devic Kingdom?

AA Metatron: *That which we term the Devic Kingdom are essentially fragmental aspects of the elemental, mineral and plant kingdoms that find conscious dynamic expression through vehicles of electromagnetic energy. Some devic forms are more advanced than others. Those of the Fae possess divine intelligence, whilst others of the devic realm are more like your animals in terms of thought patterns and group consciousness. Not all devic forms are what you may think of as positive or benevolent in nature. Some are consciousnesses sourced from electromagnetic fields, and as such both positive and negative are required to balance the electrical spectrum, you see. Some view mankind as brothers, others do not. Some are supremely loving, others are somewhat malicious, from your perspective. Yet both are electrical life forms in a manner of speaking.*

Masters, all forms of life are sacred, and it is appropriate as you grow in consciousness to attempt to understand the myriad of the great mystery. Remember, to grow in light requires eliminating fear and breaking the paradigm of limiting systems of belief.

We close by saying to you that all life originates from light, and that light originates from consciousness. From thought! Our mission on the Earth is to experience duality, to grow, to learn, to seek light. Your purpose is to experience life and to seek understanding, to make known the mystery ... and you will find that love is the vibration of all creativity, and that seeking knowledge, true knowledge, is a key element of growth.

Now, certain filters within your conscious mind are in place both by your genetic construction and by your purpose within the space-time continuum. How could you truly focus on one experience within the chosen illusion of linear duality if you were at the same time aware of being in other dimensions, living parallel 'NOW' lives. You see, it is not the purpose of the ego self to do so.

But these filters are surpassed by entering the back brain, the crystalline field! You now have the knowledge to be greater. But know that the greater knowledge is gained by seeking through the meditative mind, the God Brain, and by feeding this knowledge into the computer you call the conscious mind. Learn to operate in both!

Dear Ones, as you grow we tell you on a higher level that you are indeed already fully aware of the intricate thousands of all your soul's expansive expressions. It is through that which you call the subconscious that is the spark of God, and all of you can access it by quieting the conscious mind. Through work! Through effort! Through achieving impeccability. It is a beautiful dream, and it is as complex as Metatron's 12-dimensional cube! Make known the mystery!

I am Metatron, and You are Beloved.

… and so it is. AA Metatron via James Tyberonn

Merkabah to Merkivah ~ The Crystalline Waters of Shasta

Stewart Mineral Springs: The Amazing Crystalline Waters of Shasta
Article by James Tyberonn

The remarkable thing about the amazing crystal silica springs of Shasta is that they are such a well kept secret. The locals say they like it that way. It is not that they aren't willing to share, it is more around the fact that the rustic bohemian atmosphere of one of the most rejuvenating and healing natural springs on the planet is so available and affordable, and it is not over run by tourists. Surprisingly, it has not been bought out and turned into a high dollar five star resort spa. Interesting, because several times that has been looked at, but something seemed to intervene. It is almost like the amazing dual springs are meant to be for all people, especially the ones drawn there for spiritual refuge and healing rejuvenation.

The fact is, the springs at Stewart Mineral Springs have been revered as sacred for centuries by the Native Americans of the area. The Modoc historically described them as an area of healing, rebirth and fertility. The springs and land are still honored with White Eagle purification sweat lodges conducted by Native Americans every Saturday, and these are open to all who wish to be cleansed in that sacred way.

Tyb's Journal

It is September 2001, just days after the infamous 9/11. After having had my flight to California cancelled while waiting to board the plane at 9 A.M. on September 11th, I had through great effort managed to get to Mount Shasta on Thursday, September 14th. What a surreal week it had been.

After arriving in Sacramento, I drove my rental car to Stewart Mineral Springs for a spiritual retreat of Lightworkers. Just past Shasta City, I exited and looped under Interstate 5 and drove upward. The majestic twin peaks of the sacred mountain dominated the horizon behind me.

The road curved and changed from tarmac to pitted gravel. The entry to the springs resort was neatly couched between two ridges covered in cedar and pine, forming a serene canyon threaded by a clear gurgling mountain stream.

The rustic presentation of Stewart Mineral Springs makes it one of the most understated health spas in America. The grounds are alive with devic and nature spirits, and the springs themselves are rich with lore and legends. The energy of place seems to melt away worry and fear, and a sense of uninhibited freedom from the restraints of the mundane aspects of life seem to pervade and convert itself into liberated joy.

The Native Americans valued the springs for their healing powers, and the owners have kept the resort deliberately primitive with as few intrusions from the urban world as possible. The sprawling 37 acre site is somewhat of a random placement of quaint log cabins, tipis, residences, apartments and a restaurant, centered around the office and bathhouse spa building. The grounds are landscaped and embellished with ponds, gazebos and two wood framed bridges over the crystal clear creek. The grounds are etched with a network of footpaths and hiking trails, freshwater streams and a sparking creek pool carved out with retaining walls of natural stone. Activities revolve around hiking, sauna, massage therapy, relaxing on the sun deck and bathing in the healing waters of the amazing springs. The bathhouse entry serves as the resort office. Here one can sign up for the bath routine as well as arrange massage therapy. The bathhouse has over a dozen private rooms where the crystal spring water is heated and run into oversized tubs for soaking.

As soon as the conference meeting ended for the day in late afternoon, I went immediately for the sauna and baths. I found myself indulging deliciously in the overwhelming sense of well being that the routine of bath, sauna and cold plunge afforded me. After one complete round of bath, sauna and cold plunge, I was in land of ahhhs, quite literally over the rainbow!

On my last round of the triplex therapy, I met Stewart, a local in the sauna who was a longtime devotee of the area and part time employee. I asked him if there were any other springs in the area. I was amazed to learn that a red spring existed within a few feet of the white spring, but it was not utilized, considered useless due to its high rusty iron content. Having spent considerable time in Glastonbury, I understood immediately the rarity and the Gaic expression and meaning of a red and white spring occurring side by side. I finished my sauna and went immediately to the gazebo that housed the white spring. I spent a few moments gazing at the clear white spring and its matrix walls lined with a cushion of snowy residue, fluffy white silica. I then walked to the edge of the gazebo, and to my joy and amazement just a few feet away was a rust red pool of water draining slowly into the creek. It was unrecognized and unkempt. I jumped over the gazebo railing and put my hands into the red spring. I felt a surge of electricity flow up my arm and a communication with the conscious awareness of both springs. They were indeed a pair and meant to be recognized. I felt overwhelmed. That evening, I enthusiastically shared with the sixty plus Lightworkers attending the seminar the discovery, rarity and meaning of the dual red and white springs.

The following day, we organized a team of some twenty Light-workers with permission of the owners to clean the red spring, line it with stones and prayer ties and do honoring ceremony to acknowledge and recognize its Divine Goddess expression. As we sang songs and cleaned the red waters, to our delight, the spring began bubbling.

The white spring was covered in an acrylic bubble top that blocked the ionic plasma field that occurs above all natural springs. We removed the top, and to our joy it began bubbling profusely. The ionic field above both springs intertwined in helix.

My dear friends Michelle and Lane looked beyond the gazebo and pointed out two trees that were also intertwined around one another in a braided helix, unusual for pines. Truly this was the Cosmic wink. We were in the helixed vortex of the living springs, a fertility expression of nature where the male and female aspects were divinely coupled. My experience with the springs made an indelible impression upon my heart, and I knew I would return.

Red and White Springs of Shasta-Glastonbury

What is quite notable, rare and of extreme Gaic importance is that there are two springs on the site. One red and one white, located side by side. To my knowledge the only other location on the planet that is embellished with a red and white spring side by side is in Glastonbury, England. There, the hallowed Chalice Well red springs at the foot of the Glastonbury Tor flows side by side with the White Swan springs. The red spring of the Chalice Well is said to have been blessed by Joseph of Arimathea and Mother Mary and is renowned for its remarkable healing abilities in legend and lore.

The Druids of Glastonbury recognized the red spring as female, blood of the Divine Feminine. Legend is that the white spring was considered male, representing the white dove of Archangel Michael. In fact, the famous ley lines, the Michael and Mary lines, intertwine on the Tor and across the two springs. The area became a major pilgrimage site for centuries and still draws thousands each year.

The white silica crystalline and red iron magnetic springs of Mount Shasta have something of a similar purpose and history. The mineral waters of both are an expression of the divine male and female aspects of the living planet, embraced in a sacred tantra of regeneration. Both living springs are consciously aware within a matrix of extremely high energy, in which the veil of dimensions is melted away to allow for spiritual rebirth. Both enhance metamorphosis of the soul.

AA Metatron Channel via James Tyberonn

AA Metatron: *Greetings, I am Metatron, Lord of Light, and I greet you one and all.*

Dear Ones, we have told you that the human electromagnetic field is being stretched and pulled by the increasing energies of your planet. We have told you that in order to operate within the Quantum Crystalline field that is forthcoming within the 144 Crystalline grid, the human aura must be transformed into the greater frequencial capacity of the crystalline merkabic field.

Merkabah to Merkivah

We tell you that the geometric potential of the Merkabah is now expanding beyond the 8 pointed star tetrahedron. It is now possible to transform the frequencial Merkabah into the Merkivah, the 12 pointed star, and indeed for some of you into the 24 pointed star of Mastery. To do so requires intent, clarity and focal work. This occurs through several processes involving the trinity unification of body, mind and spirit. This involves achieving impeccability. But impeccability cannot be achieved without the purging prerequisite of trinity metamorphosis. The very build of the Ascension electromagnetics that are pulsing more and more powerfully through infinity points via the 144 Crystalline grid are expanding your own electromagnetic field. This is occurring, Masters, whether you recognize it or not. So we urge you to take heed and recognize the need to understand the mechanics of this transformation. Indeed the energy differential of Ascension electromagnetic frequency with that of your auric field must find equilibrium. That is basic physics, or perhaps metaphysics!

Ascension Imbalance

Dear Ones, how many of you are feeling the quickening of time? How many of you have trouble sleeping? You see, the inner crystalline core of the planet is now spinning faster. The ratio of the Earth's outer torque on its axis has changed in relation to the increasing inner core spin. That is effecting the time continuum and indeed all of your electromagnetic fields. Your auric fields are changing their skins. The process will intensify over the next three to four years. In the interim, your auric field can develop fissure cracks as it expands into greater frequency. This can create what may be termed 'Ascension Imbalance.' You must understand this. It can create temporary energy loss, leading to sleep disorders and chronic fatigue.

There are many tools to assist in this process. We have spoken of many of these tools: Mickel Therapy, gemstones, sacred sites, infinity points, phi cut vogel quartz, pure sonic frequencies and indeed certain living waters.

Living Waters of Shasta

And so in this gathering we will speak of the springs of Mount Shasta – indeed the male and female, the white and red springs, located in the area currently termed Stewart Mineral Springs. Now do these living conscious waters take on a greater role, now do these springs offer themselves to play a role in the Ascension of the planet. This occurs specifically through the projection of the crystalline field and the acceleration of the transformation of the human electromagnetic field into its requisite expansion. And in so doing it facilitates the transformation of the auric field into the greater frequency of the merkabic form.

Mount Shasta

We have spoken to you earlier about the special mission that Mount Shasta, the most powerful energy complex in North America, has undertaken in the Ascension – that of balancing the overly masculine energy of Atlantis with the overly feminine energy of Mu. We have told you that the essence of Shasta is primarily of the Divine Masculine, primarily Atlantean, yet located within a feminine land complex of ancient LeMuria. In so doing Shasta then projects the balance of male and female within duality. You see, the perfection, the mastery you seek cannot be achieved in either predominate male or predominate female, rather in the equilibrium, the balance of the two.

Now, Shasta is energetically articulated in more placements than just the peaks and upper valleys of the sacred mountain, indeed the energy extends throughout the enormous massif in a self contained balance of what can be uniquely termed as a Gaic chakric complex. This is a special expression, but there are other such living vortex-portal complexes that are expressed in similar patterns. These include Lake Baikal in eastern Siberia, Lake Titicaca in Bolivia-Peru, Torres del Paine in Chile, Kilamanjaro in Tanzania, Bali in Indonesia, Mount Cook in New Zealand, Tasmania in Australia, Fuji in Japan, Mount Kaleish in Tibet and Mount Araphat in Turkey to name other majors.

Sun Disc at Shasta and Atlantean Crystal Awakening

Specific to Shasta are major satellites, or energy vector expressions, within the Shasta field that significantly contribute to the overall energetic pattern of the activation of these areas into the 12 helix pattern termed the Golden Sun Disc. These include Castle Crags, Stewart Mineral Springs, Glass Mountain, Dunsmuir-Hedge Creek Falls, Sacramento Headwater Springs, Panther Meadows, Squaw Meadows, Mossbrae Falls, McCloud Falls; McArthur-Burney Falls, Black Butte and specific caves within Lava Beds National Monument. Each contributes a frequency, a singular frequency that combines to form a twelve strand helix pattern at the crater and summit. The 12 strand energy helix is what you term the Golden Sun Disc, truly an energetic pattern as opposed to a disc, yet when observed can be interpreted to be disc like in appearance.

Dear Ones, even greater expansions await you in this area. We have told you before that within the deep chasms of Mount Shasta lie one of the great Atlantean crystals. There was it placed before the waves flowed over the once great lands of Poseida, the heralded Emerald City of Atlantis. It sleeps in a deep crystalline cavern far below the vast massif of Mount Shasta. It will awaken by divine plan on the 9th of September, 2009. It was called the Crystal of Multidimensionality and is of massive proportion. It is guarded by those of Sirius B, whose cavernous base is deep below the ridges of Stewart Mineral Springs. We will speak of this in the months to come.

Activation of the Shasta Spring Crystalline and Magnetic Springs

Now, the springs at Stewart Mineral are activating into their unique role that extends to all who visit them. We have spoken about previously and now reemphasize the requisite expansion of the human auric field that is occurring in the Ascension. The planetary pulse is speeding up, and Dear Ones, it is affecting you in myriad ways. In a very valid sense what is occurring through the ascension energies is stretching your frequencial capacities. As such you are required to grow a new expanded auric field, much like a snake that annually outgrows its skin and goes through a susceptible phase of fitting into a new one that can better encompass the larger body. Do you understand? Before the new one can become resilient, the old one is stretched and cracks, and during that transition there is a phase of metamorphosis that contains within it certain vulnerabilities until the new one is completed.

The unique mineralogy and location of the crystalline and magnetic springs of the Shasta field are activated to assist in the adjustment, healing and expansion of the human electromagnetic field. The rare silica content of the white spring is one key. The spring is in essence crystalline in frequency, silica being the base component of quartz, you see.

We have made the channel aware of the special role of these springs. He was urged to prompt an analysis of the two waters, both the red magnetic and white electrocrystalline waters.

Question to Metatron: I have always felt since discovering that the red spring was not being utilized or recognized that it was also to be used? How can the red, female spring be better utilized?

AA Metatron: *The two springs should be used in tandem, are meant to be used in tandem. Optimally the white electrocrystalline waters should be used as the initial bath. This provides a deep purification as well as electrocrystalline expansion. The detoxification achieved from the white spring occurs both within the body physical, glandular system and the etheric field. Emotional releases of blocked energy are accelerated, especially when the baths are used three to four days in sequence.*

And indeed, the red iron magnetic spring of the female frequency should be the final bath after completing two to three 'rounds' with the crystal spring. The red spring then as the final soak provides a soothing magnetic rebalancing effect that synergizes and completes the dual use of both springs.

These springs, as we have said, have taken on a new role; an activation is occurring. They will uniquely be of benefit to the physical, emotional and electromagnetic fields of the body. This utilization will accelerate and assist in completing the expansion of the human electromagnetic field and help what is termed 'Ascension Imbalance.' You see, the requisite purging of emotional blockages will be accelerated within the energy of the Shasta vortex and further accelerated by bathing in these living waters. One is electrical, the other magnetic; both carry the crystalline energy. We will tell you that the red spring also offers many unique healings; those with arthritis will benefit immensely from the magnetic therapy of these waters.

Question to Metatron: The red spring is currently not set up for use in the bath house. Is there an alternative method in the interim for this to be used?

AA Metatron: *Despite the unfortunate non use, non recognition of the red magnetic spring, the benefit of the white spring, the electrocrystalline spring, is considerable. As both the red and white springs overflow into the stream used for the cold plunge, the waters of the flowing creek can offer some of the qualities that otherwise would be obtained by bathing in the red spring. It can suffice but does not offer the full benefit, you see, of the red spring. But in the now, the plunge into the mineral rich creek after the sauna is an important aspect of sealing the energetic and frequencial expansion brought by the crystalline spring. The flowing waters themselves offer a hydroelectric energy that is vibrant. The flow of Park Creek is in fact a hydro ley line, rich in akash, adamantine energy. A hydro line is very similar in benefit to that of a ley line. Around the two springs is the epicenter of a regenerative vortex complex that fills the small canyon of the grounds. Well to spend time in prayer and meditation at the well source of the two springs, honoring both. This can be done before and after the bathing, sauna and plunge routine at the gazebo source of the springs.*

In terms of utilization of the magnetic red spring for bathing therapy, the seed has been planted. The spirit of place has pulled in enlightened caretakers who will recognize the call, and others are being drawn to assist. Optimally the red spring must be truly acknowledged and then utilized in the format we have described. In truth, neither of the two are fully recognized or honored for their very special offerings and rare occurrence. The two combined represent a living, conscious field of regeneration, of fertility, of the combining and coupling of the male and female expressions of Gaia within a unique field of rebirth.

The recognition is beginning to take hold. After our prompting an alter has been placed, and several times has the red spring been cleaned, lined in stones, and prayer ties have been placed at its tangent perimeter.

The concept of recognizing and honoring the two springs at their source is vitally important. Both the female and the male component are to be recognized and allowed to helix. Both the electric and magnetic should be used to fully optimize the unique benefits being offered.

Ultimately a new gazebo, a sacred geometric patio of sorts, should be constructed that encompasses both springs within the form of the Vesica Pisces. On the outer circle of one should be the raised pool walled in opaque snow-quartz stone for the white spring, in the other outer circle the raised wellhead walled in red stone or granite to contain the red spring. Overflow drainage should be designed in each pool to allow for a water overflow to helix into a common pool that blends the waters of both. Neither spring should be covered in solid hardcover, as is currently utilized to prevent obstructing refuse, coins or offerings being placed in the white spring. Better to use a thin gauge stainless steel mesh or amply perforated acrylic, which allows the ionic plasma produced by each spring to flow freely upward. This then allows the natural entwining of the energy of both springs to join and helix. It is the natural pattern, the desired expression of both.

We will add that taking water from the red spring and pouring it over the body after the process would be beneficial although cumbersome within the current set up. Optimal use is the full bathing.

Question to Metatron: Because the lands were once part of the Native American sacred grounds, some have suggested there is an energy conflict upon the land of Stewart Mineral Springs. Is that the case?

AA Metatron: *There is no conflict on the land. None!*

The lands, because of the flowing waters and sacred energy of the springs, keep themselves cleansed of conflicting thought form energies. In regards to human conflict around use of the land or ownership of the land, these issues are quickly disseminated by the energy of place. The purification sweatlodges that are regularly conducted also imbue a pure energy into the land. But then again, so does the energy of those savoring the tranquility and rejuvenation of the spa imbue itself benevolently to the land. The land, the spirit of place, is quite aware of itself and of its role and purpose.

Indeed there exist pockets of what may be termed raw electromagnetic imbalance within certain areas of the land, especially on the hillsides near some of the cabins. This occurs in most vortex sites of full spectrum energy. In electromagnetic terms, the positive cannot exist without the negative. Yet these pockets can be made more congruous with the harmonics of the land with intent, and the areas along the moving waters and especially the springs glow in great positive light.

But although the lands are kept conflict free in human terms, there remains the need to acknowledge and utilize both springs. So we will say that the land of the springs is offering itself for greater utility and is calling to those who will hear, so that the optimal purpose and need for honoring recognition be acted upon. Until this is completed the land is not fulfilled, and indeed this may be what some interpret as conflicting energies. But this is an interpretation, a misinterpretation. The springs call for unified acknowledgement in order to complete their role, you see, and that energy will be felt until the utilization of the land, of both springs is fulfilled. So the land is not in conflict, rather in a phase of incomplete recognition. Do you see the difference?

No, we will add that it is essential that the owners and caretakers of the springs maintain and follow the requisite purpose and stewardship of the lands and springs. Not to do so will result in imbalance.

We will say that in current times it is difficult to maintain the required overhead and necessary financial returns to keep the healing spa functional. The waters are meant to be used; the grounds are meant to be shared and available to all. The temptation to over commercialize and convert to a private high dollar resort must also continue to be avoided. These waters must not be restricted. Indeed the owners have accepted the guidance of deliberate underdevelopment, yet it is important that the red spring become a recognized and utilized part of the spiritual and physical healing therapy. To do so will require financial energy. The best means is to develop a nonprofit fund to raise monies for the 'Vesica Pisces Alter' housing both springs as described. The intent should be expressed for a landscape architect and attorney to volunteer assistance.

The installment of the bath facility to utilize the red spring for bathing will follow suit. Indeed will it bring greater financial reward to the facility. Indeed a reasonable abundance exchange for the access is appropriate.

Question to Metatron: Can you expand on how the crystalline spring assists in the Ascension?

AA Metatron: *As we have said, the spring assists by providing a unique crystalline vibration that is synergistically imbued with akash and a rich mineralogical alchemy, which when absorbed within the human body and field is capable of detoxifying, contracting and then expanding the field into greater crystalline capacity.*

By the combination of female-male, of electric to magnetic is the balance obtained that greatly benefits the chakric and glandular systems. The waters themselves, the energy of place, the spirit of place assist in the removal of blockage, and the revitalized energy field is expanded and renewed within crystalline balance. Do you understand? This is one of the very rare silica waters on the planet that can be used in this manner.

Now, the planet contains many sources of springs. Not all of these are within infinity points such as Shasta. Not all springs are truly self aware. Living springs carry great electrical energy, being pulsed through the Earth's strata somewhat in a jacuzzi manifold fashion that imbues great energy and minerals into them. Such springs are very beneficial to the body and emotional fields.

Just as many now recognize that the consumption of living, raw foods allows the body to take in energy fields from the living plants, so does the same occur when the body is soaked and bathed within living electromagnetic waters.

A similar capacity exists within Banff National Park in Canada, but there the waters are treated and diluted such that the potency is somewhat lessened though still beneficial. Because of the more natural aspect of the twinned springs at SMS, and because of their location within the field of Shasta, they are far more potent. Living springs occur in every part of the globe. These are especially potent in New Zealand at Silica Springs Tongariro, the Yucatan area of Mexico, Montezuma's well in Arizona, Caldas Novas in southern Brazil, Termas de Araxá in Mineas Gerais, Hungary, Bath, England , Italy, Turkey, Japan, Peru and in Hot Springs, Arkansas.

But we must add it is not simply the springs, it is also the consciousness of the living springs themselves and where they are located. Not all springs carry potent aware life force. Not all living springs offer themselves for mankind. Those of Yellowstone and Lassen serve different purposes for the Earth itself. Accordingly Stewart Mineral Springs are quite unique and of the highest benefit for mankind at this time.

Now, at Stewart Springs, the routine of utilization should occur three to four times within a seven day period. Well to do this at least twice per year, optimally once a month. The sequence of bath, sauna and cold plunge is correct. It can be repeated as often during each day as wished. Because of the high alkalinity, pH, the user must discern when it is time to stop lest skin irritation occur. We will add that utilization of living springs on a weekly, monthly or even quarterly basis can extend the life span of the human. Any usage is of benefit.

Question to Metatron: Are these springs aligned to the Chalice Well and Swan Springs in Glastonbury, England**?**

AA Metatron: *Absolutely. Not just the springs, but Shasta is aligned in full to the greater complex of Glastonbury to include the Chalice Wells, Tor, Wearyall Hill and Abbey. Glastonbury embodies the Christ Consciousness of the true King Arthur, indeed of Camelot. Shasta expresses the Christ Consciousness of that termed the White Eagle and indeed of the Ascended Masters and Saint Germain.*

The red and white springs of both locales have a very vibrant axial tonal connection. Considerable exchange occurs. You may note that the red spring is recognized and utilized in Glastonbury while the white spring, male, is not. Paradoxically the white spring is utilized in Shasta, and the female red spring is not. Does this help you to ascertain why they are especially in harmonic communication?

The channel has asked if the springs in Shasta are meant to be drank, as are the two in Glastonbury. The answer is no. One need only review the chemical analysis to understand why. The springs in Shasta have a high pH, approaching 11 in the undiluted state. This is excellent for the detoxification and expansion of the auric field in bathing yet far too alkaline to consume orally. Additionally the waters of the springs in Shasta are somewhat colloidal in nature, and the solid content of the suspended particulate is too high for consumption. Thus the waters of Glastonbury are for drinking; the sister-brother springs in Shasta are for bathing. Both are greatly beneficial for the body and for expansion of the electromagnetic field.

We will add that the charged springs of the Sacramento headwaters are extremely beneficial for drinking, as are the spring waters of Panther Meadows. Yet in your current environs, both should be filtered before consumption. A portable filter such as the Brida type can be used and does not decrease the life force within the waters. However, the electromagnetic energies within the waters will decrease quickly in stagnation and storage. Best to drink immediately after filtering and within one week if taken to the home.

Closing

And so Dear Ones, as the resonant magnetics and electrics of your Earth are increasing by design, so do these upsurges effect the energy of your human EMF. While some of you will experience this differential with comparative ease, all of you must be aware of its effects. The differential can create temporary fissure cracks in the Human EMF and can lead to energetic bleeding, termed Ascension Imbalance. The expansion and recreation of a larger merkabic field is a requisite metamorphosis of the auric field to the Merkabah, and indeed of the Merkabah to the MERKIVAH. The Merkivah expands to the 12 pointed star, the 24 pointed star and in time well beyond. The fabric of your universe, the matrix of higher dimension, is sacred geometry within a crystalline field. You cannot, will not experience its heights until you become crystalline. It will not happen by itself.

Dear Ones, the electromagnetic field of your physical body must be understood and maintained. It interfaces much of what you term the soul memory – emotions, feelings and spirit – to the physical experience.

All of you are in a crucible that will offer you the opportunity to purge the imbalance, remove the obstacles and allow you to expand beautifully into the Quantum Crystalline field within the 144 grid. You must consciously work for this! That is why you are here!

Please understand that if you are feeling irritable, out of balance, fatigued or overwhelmed, you are not alone. The duality of your plane is a great teacher, a growth cycle that you have waited eons to complete. Understand what is occurring, and see this as a great opportunity. A GREAT OPPORTUNITY. Seize it! There are tools to ease the process. The living waters of Shasta are one such specialized source of assistance. Indeed they can assist you in achieving the crystalline MERKIVAH.

Love yourselves, Masters, and love one another. Be gentle with self, and be kind to one another. Hear these words. I am Metatron, and I share these messages of Light with all of you. You are not alone, we know each of you by name, and you are Dearly Beloved!

…and so it is. AA Metatron via James Tyberonn

Grand Canyon ~ Rebirth in the Womb of Gaia

Article by James Tyberonn

The Grand Canyon is listed as one of the seven natural wonders of the world and is at the top of the list of America's most beautiful National Parks. Yet surprisingly, it is rarely mentioned as a sacred site or powernode. I say surprisingly, because it is absolutely one of the most potent energy sources on the planet. The Grand Canyon is not only one of the wonders of the world but also a sacred site of epic proportion, a mega powernode of the Ascension. It is the womb of Gaia. The Canyon is chock full of massive energy fields, vortexes, and portal stargate complexes. It is also a dimensional doorway, an entry point to the hollow inner world of the earth.

I have had the opportunity to raft the first 200 mile stretch of the 277 mile canyon on two occasions, and both excursions profoundly effected me on every level. Truly, you cannot experience the full magic of the Canyon unless you raft its length. Every stretch and turn takes one deeper into the earth and in a very real sense deeper into one's self.

The Canyon is a testament to the National Park Service, whose rangers and employees assigned to the Grand Canyon are among its most fervent guardians. A limited number of people are able to raft the stretch each year, as volumes are controlled for environmental purposes. It is in fact maintained so tidily that the Guides tell each rafter on the outset, "The only thing we leave in the Canyon is our footprints!" Wait lists for the eclectic experience are often two to three years. Even after a century of tourism, the spectacular pristine beauty of this uncorrupted desert canyon oasis yet remains unspoiled and uninhabited, enticing world weary travelers to seek deep peace and contemplation away from the maddening crowd, as indeed it did for this writer.

Tyb's Journal

I rafted 200 miles down the Grand Canyon the first time in 1996. The beauty left me in awe. At the entry point at Lee's Ferry the Canyon walls are about 20 meters high, and the river is a translucent shimmering aquamarine. The Canyon walls change day by day from sandstone orange to deep red, back dropped by majestic temples and castles. The waters go from green ripples to café crème torrents, roaring into class five rapids.

Each day our river guides lead us on hikes to incredibly beautiful cascades hidden at the back of chasms smoothly carved in artistic curves by the eons of erosion. Crystal clear waterfalls await us in faerie dens amidst hanging gardens of green fern and yellow bracken growing in rock wall inclusions. Soon the walls are over 5,000 feet high, and we enter mysterious lunarscapes, enormous temples and stark cliff faces that remind me of Yosemite's towering 'El Capitan.' We enter a red stone cave with golden sand floors, large enough to house a full symphony orchestra, which indeed it did several years ago. We hike to places of stunning beauty like Elve's Chasm, Deer Creek Canyon and Havasu Falls. Places so surreal, so mystical, so pristine, you expect to see a T-Rex round the corner ala Jurassic Park any moment. We spend precious time in areas so remarkably special that each person seems to take time to cognitively record every detail. I literally began to look at my watch each quarter hour and dread the time set to leave and hike back to the waiting rafts. We visited the crystalline blue Little Colorado, its turquoise hue surreal and alive, and its floor tiled in massive clamshells of white calcium carbonate.

At night we watch amazing sunsets of blazing red morph into burnt orange and magenta. As the sun falls beyond the horizon, the sky becomes a dark velveteen canopy, lit by millions of twinkling stars. The humidity free air offers an amplified starscape that seemingly swims and floats in deep spirals. The rush of the river teams with the bass rhythmic croaking of frogs to create a sonic cocktail that mesmerizes us even further as we sit around campfires, jubilant and pensive. We deeply savor the moment.

It ended ever too quickly, and I vowed to return as the helicopter lifted me above the rim. I did return three years later and experienced an even deeper metamorphis. I grew to love the Canyon even more. I expanded my beliefs to understand that the Canyon is a Living Cathedral that indelibly is equipped to transform and vitalize each person who visits its spectacular space.

Sacred Space

The Native Americans who live in the Canyon had formed their own vivid picturesque conceptions of the genesis of the spectacular chasm long before European explorers and modern geologists entered the area. Their descriptive narrations were expressed as spiritual metaphors and poetic myths recounted by shaman, elders and storytellers from generation to generation. To the Havasupai, Hopi, Navajo, Ute, Southern Paiute and the Hualapai, Mother Earth, Father Sky and Grandfather Sun are alive, consisting of natural and spiritual forces that are inextricably joined as part of 'Great Mystery,' *Taiowa*, the Creator God.

Each aspect of nature and of the Canyon has a spirit. The great river that formed the Canyon gorge led to the 'promised land,' and the life giving crystalline springs had spirits to be placated and honored by offerings of sage, corn and tobacco. The animals that the Indian hunted, the avions that flew above his dwelling, the snakes that slithered across his trail were man's brothers, serving as providers, guardians, messengers and spiritual teachers.

Native lore states that the Canyon was the result of the Great Flood that had drowned the previous 3rd world, a world that had forgotten the path of truth. The Grand Canyon is said to be the Eden Point, the beginning of the 4th, our present world. The Hopi cosmology is clear that the Grand Canyon was the *Sipapu*, the 'place of emergence' from whence the Hopi had emerged from underground after the flood had destroyed the 3rd world. Several inner world 'Sipapu' entry points are said to be located on Hopi land in the Canyon, one of which is honored in ceremony as the dwelling of the ancients. Appearing as an opening in a rounded rock mound, it is a sacred place of pilgrimage for the Hopi, located at the bottom of the Canyon of the Little Colorado River just above its joining point with the Colorado River. It is very hallowed and off limits to all but the Hopi people. The lore further claimed that the Hopi were assisted by 'insect people' who live in the inner world caverns, humanoids who had stick like arms and legs, slightly arched backs, and blue toned skin.

Native Americans of the Canyon revered the movements of the stars through what they termed Star Wisdom. For the Hopi, beyond the Mother Earth was Star Nation. Their legends speak of Star People who visited through holes in the sky located within the Canyon, a clear reference to what we consider portals, stargates. The Hopi and Navajo consider themselves direct descendents of the Pleiades, whom they call the *Chuhukon* and *Delyahey*, respectively. Their legends speak of White Star Children, who helped them as Gods. The Pleiadeans are said to be Nordic in appearance.

Mysterious Egyptian Cave in Grand Canyon

The April 5, 1909 edition of the Phoenix Arizona Gazette newspaper featured a front page article entitled, *"Explorations in Grand Canyon: Remarkable Finds Indicate Ancient Egyptian People Migrated from Orient."*

The lengthy and detailed article stated that the expedition was financed by the Smithsonian Institute and discovered artifacts that could potentially change our understanding of history. Inside a massive cavern with intricate passages and rooms that Kincaid described as "hewn in solid rock by human hands" were found tablets bearing hieroglyphics, copper weapons, bronze tools, huge statues that he described as appearing like Egyptian and Tibetan deities, as well as over 500 mummies. Although highly intriguing, the story has been officially disavowed. The Smithsonian denies any knowledge of the discovery. Researcher, author and explorer David Hatcher Childress writes, "While it cannot be entirely discounted that the entire story is an elaborate newspaper hoax, the fact that it was on the front page of a prominent newspaper named the prestigious Smithsonian Institution and gave a highly descriptive and detailed story that went on for several pages, lends a great deal to its credibility. It is hard to believe such a story could have come out of thin air." The article is in the newspaper's archives, both in original print and microfilm, and can be easily found on the internet.

To add more to the conversation, renowned author Drunvalo Melchizedek refers to the 'Egyptian' aspect of the Grand Canyon in volume II of *"The Ancient Secret of the Flower of Life."* On page 302 of his book, he describes the adventure of two backpackers who hiked into the backcountry of the Grand Canyon toward the pyramidal mountain called the Temple of Isis. After an arduous trek, they claimed to have come upon what appeared to be a large impressive man made pyramid, constructed of natural stones. It showed signs of erosion but still held the distinctive pyramidal shape and definitely did not appear to be a natural formation. After a thorough inspection, photos were taken, and they continued trekking to the amazing formation of the Temple of Isis.

Upon arrival to the Isis Temple formation, they climbed up some feet, and from that vantage point several interesting cave entrances came in view. They observed that the grotto entrances were cemented and blasted shut. Both climbed upward to the most prominent cave opening, which had an arched entry. They reported that the entrance showed signs of chiseling, and the arc appeared man made. Just above the entry they observed a six foot diameter circular pattern with inner designs clearly chiseled into the ceiling. Just beyond, the cave had been completely filled in with rock and stone rubble, appearing to have been blasted shut, making further entry quite impossible. This fascinating and very interesting experience was recounted directly to the author, Drunvalo Melchizedek, a man of impeccable integrity.

Comedian Steve Martin used to include a song he wrote called *"King Tut"* in his Saturday Night Live routine, the opening line of which goes, "King Tut, King Tut, he came from Arizona and he moved to Babylonia," … maybe he knows something we don't!!

To add a bit more intrigue, researcher Robert Morningside listed over 20 striking similarities regarding cosmology beliefs among the Babylonian Sumerians and the Hopi. Both believe mankind was created by a woman God, called *KA*. The Hopi call the Sun Creator *Taiowa*, the Sumerians call him *Ta-Ea*. The Hopi and Sumerians use the same name, *Pahana*, in reference to a Divine Star being who will return one day to assist mankind.

From a personal perspective, I have become quite aware of the extensive tunnels that exist in the inner earth in numerous locations that I have visited across the planet, particularly in Peru,

Bolivia, the Himalayas and Egypt. I have no doubt these exist in the Grand Canyon as well as in the San Francisco Peaks near Flagstaff Arizona, the latter of which I vividly experienced in lucid dream during my fourth vision quest in the volcanic calderon atop the San Francisco Peaks. Whether the Canyon Cave story is factual or not, the Canyon has a mysterious aura in some sections where other world dimensional aspects seem to emanate, and these can be easily sensed by empaths.

Interestingly, during my two rafting trips through the length of the Canyon, I experienced very clear sensations of an 'Egyptian' resonance and connection within certain areas of the canyon and had a powerful visionary experience of Lord Buddha in another. This was prior to my having any knowledge of the 1909 article of the infamous Egyptian cave. It is quite intriguing that many locations in the Canyon have Egyptian names and Buddhist names. The areas around Ninety-four Mile Creek and Trinity Creek are embellished with stunning formations named the Tower of Ra, Horus Temple, Osiris Temple, Isis Temple, Cheops Pyramid and the Tower of Set. Further downriver near the Havasupai Tribal lands within the Canyon are names such as the Buddha Cloister, Buddha Temple, Manu Temple and Shiva Temple.

Tibetan-Grand Canyon Hopi Connection

> *"When the iron eagle flies and horses run on wheels, the Tibetan people will be scattered over the earth and the dharma will return to the land of the red man."*
> Tibetan Lama Padmasambhava, 8th c (aka Guru Rinpoche)

Several years ago, I attended a Native American modality, 'Rainbow Tribe Earthdance,' along the edge of the Grand Canyon. A Havasupai Elder attended the ceremony. Before we were introduced, I thought he was a Tibetan Llama. The oriental characteristics were quite striking.

Beyond the similar physicality, there are other noted comparisons between the Grand Canyon Indigenous and the Tibetans. Several articles have been published that offer evidence that Tibetans and Native Americans share a common ancestry. Author Frank Waters observed in "The Book of Hopi" that the indigenous Zuoi Shalako or snake dance symbolically mirrored the Tibetan ceremony of the Journey of the Dead. He also noted that both cultures recognized and mapped the human chakric systems long before Europeans inhabited the Americas. Other commonalities listed include cosmology, similar jewelry designs and inlays, use of silver, turquoise and coral, nearly identical patterns of woven textiles and the same distinctive manner for braiding hair.

Perhaps most complelling is explained by author Daniel Pionchbeck in his book, "2012: The Return of Quetzalcoatl." He states that specific words in the Hopi and Tibetan language are phoentically exact with antithetical mreanings. The Hopi word for day, nyma, is the Tibetan word for night. The word for moon in Tibetan, dawa, is the phoenetically exact Hopi word for sun. If you were to drill a hole straight from the Hopiland reservation directly through the Earth, you would come out in Tibet. Interestingly, these two 'reversed' words are quite precise in describing simultaneous planetary time. In other words, when it is day in the Canyon – nyma – it is night time in Tibet – nyma. When the moon, dawa, is out in Tibet, the Hopi see the sun – dawa.

In ancient days could they have been able to instantly transport within the dimensional tunnels said to connect them? Maybe the Canyon will one day reveal all of its secrets, then again maybe it already has! Truly, it is a place of great mysteries that call for further exploration. One senses whispers of answers within its depths … or is it just the wind?

Closing

At the beginning of the rafting journey, our river guides told us that the only thing we would leave in the Canyon would be our footprints. They were wrong; we all left a part of ourselves within its majestic depths, especially our hearts.

Archangel Metatron Channel via James Tyberonn

Greetings, I am Metatron, Lord of Light. And so we come to you at this time in a moment of wisdom, your wisdom and the collective faith of all wisdom. Dear Ones, it is wisdom you seek and wisdom we share. It is the role of Spirit to offer you perspective and possibility. It is your role to discern that which resonates and apply it to your lives as you seek your own inner truth, your own path to SELF Divinity.

And so we speak to this gathering on the Grand Canyon, truly one of the most powerful energetics of the planet. A spectacular lotus energy, one that is of great intensity yet expressed within a magnificent harmonic that allows for a gentle receival by those fortunate enough to explore its depths. Within the rims and depths of the vast isolation of the great chasm are some of the most pristine and benevolent energy fields of the living Earth.

Now, the Earth is a living, conscious being, and Gaia is her sentience. There are certain places within and without your Earth that are living libraries, information repositories, we would say. These are not simply registers of the past, indeed they also contain information regarding the present, the future and all of their probabilities and parallels. These living libraries only occur in special environments, environs that perfectly synergize all of the elements: earth, air, fire, water and ether. Within these frequencies, indeed one also has the ability to learn immensely about one's self. Such locales are indeed rare. The Grand Canyon is one of these, the largest and most complex library on your planet. In a true sense, the rafting through these fields in the Canyon core is a rebirthing process, regeneration within the chakric womb of Gaia.

You see, ether, the fifth element, generates a panacea, an essence we term akash or adamantine from the inner realms of the planet. Akash occurs in great abundance in the Canyon and provides a bounty of benefits beyond its tell tale energy of well being. Etheric akash is indeed the rejuvenative elixir of life and is merged synergistically within the ionic plasma of the dynamic river. Thus it is, in your vernacular, the fountain of youth, a dynamic regenerative force!

As such, visiting the Grand Canyon in any location above or below is a beneficial experience in terms of its effect on the human energy field. A vast energy is emitted and amplified within its substratum, condensing and rising to the rim where it spills over and feeds an enormous vortex that circulates for hundreds of miles.

Yet it is only within its depths that an incredibly more powerful energy accumulates. These energy pools in the base of the Canyon are quite unique in their individual identity and expression, because above the rim all of the individual aspects are somewhat blended into the massive spinning vortex that contains them all. The floor of the Canyon, the portions along the bottom between the towering walls, are ten fold more powerful than the upper rim of the Canyon, containing concentrated energies that are in a continual state of regeneration. You see, electromagnetic energy, the basic source of ley energy, flows symmetrically within the pancaked layer of each stratum. When that stratum is sliced open at the cliffs and walls, that energy flows into and amasses within the openings of the Canyon where it is in boxed containment. It is charged further with the energy of all five elemental forces and uniquely so with the addition of the most rare element, akash ... etheric life force.

The ionic field created by the brisk flow of the copper rich river amalgamates with akash. This creates a uniquely gentle stabilizing force that allows for the human electromagnetic field to interface extremely expansive energies without fissuring under acute pressure differential. As a result, the human EMF is so enhanced that it is quite unlikely for one to experience negativity within this kinetic cornucopia and quite likely to experience a deep epiphany. As such, one's auric field is expanded and dimensionally redefined with a greater paradigm.

One would not expect such a 'raw' landscape to provide such a nurturing embrace, yet truly the Canyon is Gaia's womb. Indeed nurturing occurs, a true rebirth is provided by the serendipitous harmony of its stark dichotomy. Is that not a paradox?

Those of senior age, those with slight infirmities who may have doubted their veracity in signing on for such an arduous adventure soon find themselves bursting with robust vitality ... rejuvenated! Accordingly, myriad benefits are enhanced, including a revitalization of the body physical, indeed the life force, the libido, is greatly augmented. A clearing of the emotional field, a realignment of the chakras, mental exuberance and a vast expansion of the auric field will occur, indeed must occur, in these natural forces.

We tell you that within the walls of this creative crucible, this planetary womb is contained every telluric energy resonance on the planet in a magnificent umbilical feed. There is a resonant harmonization that sings every vibrational tone into a chorus of orchestrated symphony. How can such vibratory energetics not be extremely healing to the mind-body and spirit of all within this space? This is why the Canyon has been recognized by many, including those of Lemuria and Atlantis as a frequencial 'Garden of Eden.'

Planetary Libraries and Harmonic Oscillation

Layers of earth as they amass on the planets surface, especially crystalline rock, record every vibration within their fields. That is how and why certain areas of perfect elemental symmetry are Planetary Libraries.

We have spoken about the Law of Harmonic Oscillation before, the aspect of similar mineralogical frequencies naturally attracting and activating one another. There is another aspect to this that involves vibratory resonance within the time space continuum which includes geological ages. Specific surface exposed strata from the Precambrian class, for example, will naturally align with other surface exposed Precambrian strata, not just because of similar mineral construct but also because of similar time sequence vibrational imprints. Vibrational imprints of a specific time era will include the magnetics of the planets, stars and cosmic alignment frequencies as well as mankind's vibrational forces during the vast epochs of surface rock. Do you understand?

Now, as you may have perceived, the harmonic oscillation of certain of these areas is in alignment with Egypt, the Himalayas and Tibet. This trait is amplified further in igneous and metamorphic forms, because these formations are more crystalline in composition and thus produce a greater electromagnetic field, you see? It is the Harmonic Oscillation process that aligns the

Grand Canyon to Egypt and Tibet. We will speak more of this later.

Inner Earth

The Grand Canyon indeed has many entrances into the chasms of the inner world. As we have told you, your Earth is truly not spherical. It is flattened at both poles. Your planet has five distinct shells (not four) within its structure, one inside the other. Between the layers of the outer three are great chasms, some of which have flowing rivers and have illumination from ionic fields, much like the northern light phenomena that you term auroras. One of the largest of these stretches below the Grand Canyon for an expanse of over 800 miles in length, by your measurement, and is over a mile in height. The Atlanteans in the peak of their technological golden age were well aware of the peoples inhabiting these chasms, and indeed they interfaced extensively with them in the construct of the extensive tunnel system.

Do you have questions?

Question to Metatron: Yes, thank you, I have several. First, you have spoken about the energy fields found along the 277 miles of the Colorado River within the Grand Canyon. Which areas are most potent?

AA Metatron: *There are dozens, and so to answer your question regarding which are most potent, we assume you mean which are most beneficial, yes? And so we answer that the degree and priority of assistance depends to some extent on the person. However, it must be said that the course of the Canyon along the river stages the rafters in a regimented common process. It is meant to be so. The rebirthing would otherwise not occur in the correct sequence. Now, within that regiment, the* *seeker will receive what is most needed, and that will of course vary according to the individual's level of light quotient, intent, attitude and needs. In terms of field potency, the area of the Vishnu schist formation is extremely powerful, such that it is the primary generator of the highest frequency vortex and the center of the higher dimensional portal. The area of Blacktail Canyon is one of the most powerful portal stargates on the planet, in an extremely congruous alignment to the Pleiades. Those capable of raising their frequency to higher levels can experience what you term as astral travel within this area far more easily than in other locations. This because by the time the Vishnu Schist of Blacktail Canyon is reached, after threes days of staged expansion and balancing on the river, the auric field is cleansed and is thus in a crystalline merkabic state. The mind is clear, the body is robust, and therefore the body-mind-soul trinity is in a rare state of unified symmetry and thus clarity. Do you see?*

Now, the akash elixir is an imperative ingredient in this process. Without its abundant presence the rebirthing would not be possible to all, because the intensity of the telluric forces would be too great. The akash is in greater abundance in the Grand Canyon than any other locale on your planet. It is in fact being generated in greater abundance now as your planet begins its transformation. It is generated through areas of tectonic stress, particularly around certain volcanoes such as Mount Shasta and Hawaii's Kilauea. But nowhere is it generated in such quantity and held so purely in place as occurs in the Canyon. This essence is extremely beneficial to the human EMF and its emotional and mental components. It aids in synergizing them into perfect symmetry and balance. A very rare, very aligned state of well being occurs, as such.

Now, in addition to the metamorphically crystalline Vishnu schist, the Zoroaster Granite within the Canyon is also extremely potent.

The silicate component within the granitic field acts in unison with the Vishnu formation. The granite serves as the receiver and the metamorphic schist as the balancing anchor for a tremendous portal within this area. Most of the extraterrestrial craft that enter and work beneath the Canyon enter through this portal. Indeed it is a stargate to the Pleiades and an entry portal to many from extraterrestrial realms. It is also a time gate.

You ask about other areas of benefit. The Havasu area along the Little Colorado Creek is also extremely beneficial in terms of physical and emotional healing. It is teeming with those of the Fae, the Faerie Kingdom and the Devic water sprites. The Fae within this location are of an extremely tangible and harmonic resonance, easily discerned, and they add that element of lush richness to the plant life. Indeed this rich devic energy expresses itself in an uplifting joyful vibration, a sense of gentle well being to all who enter. It is the perfect afterglow to the intensity of the Vishnu energy. The blue green waters of Havasu are embellished with unique arrays of fortifying minerals, including copper, calcium carbonate and magnesium sulfate, what you term Epsom salts. The color of the water is in itself is very healing very soothing, its mineral content is suspended and dissolved to an almost colloidal liquid and allows for optimal absorption into the body physical.

Likewise the area of Deer Creek Canyon is extremely healing, containing a plethora of devics of the elfin nature, joyful in expression. All of these positive beings add to the vibration of these pristine areas. The Redwall Cavern site is also quite noteworthy. It exudes a sonic vibratory resonance that is most conducive to deep meditation and out of body travel, that you term astral travel experience. The sands within this erosion chamber are also extremely vibrant and mineral rich. Well for one to cover the body in this sand for a 30 to 45 minute period. Quite enriching as well as detoxifying. These are but a few; the Canyon indeed is a cornucopia of beneficial energies, quite unique to this area.

Does this answer satisfy you?

Question to Metatron: Yes, thank you. There is alleged information about a supposed discovery of an Egyptian or Tibetan cave in the Canyon. Did the discovery of such a cave occur in 1909 in the Grand Canyon?

AA Metatron: *Absolutely! It did indeed occur. A cave was discovered in early 1909 and revisited several weeks later with a larger expedition, which included anthropologists and assistants. It was, however, neither Tibetan nor Egyptian, rather Atlantean. It was part of the extensive labyrinth of tunnels built by the Atlanteans that circumnavigated the planet. This was done in an age of great technological expertise and was built with technology primarily from Sirius B.*

As we have told you in previous sessions, your planet, while not hollow per se, contains great open chasms, some of which are massive and indeed inhabited by the blue skinned humans who 'descended' into the inner earth from Lemuria. They are indeed human, but at a far more advanced cycle in their development. As such they operate in the 4th through 7th dimensions. Several groups of extraterrestrials also work within these chasms.

Question to Metatron: It seems surprising that in 1909 the government would have sought to hide the incident. Was this in fact covered up, and if so why?

AA Metatron: *The discovery of the opening was 'covered up,' in your terms, only after the third entry, the expedition that included the Smithsonian team. A catastrophe occurred that led* *to the death of most of the team; as a result the event was suppressed, and the entries were sealed during a third and fourth visit specifically for that purpose in 1911. Only the first two expeditions actually explored the cave.*

The initial discovery by the one called Kinkaid was rather uneventful, other than the tremendous excitement of the explorer who saw the artifacts and the Atlantean hieroglyphics and statues, and he recognized them as being from an advanced civilization. He did discover statues, tools and mummies, and he drew diagrams and maps. Kinkaid spent two to three days exploring multiple surface chambers and departed with his drawings, descriptions and indeed photographs. He departed after four days because of the lack of nearby water and the arduous, treacherous climb and distance from the river.

The second entry occurred within eight weeks of the discovery and included a hastily gathered team of eight members, including two military guards and two academics. During their extensive explorations and mappings, they descended further and were drawn to an amber light pulsing in a circular chamber containing a massive hyper dimensional generating complex used for transporting within the tunnel system. Most of these are dysfunctional relics, but some are maintained and undulate into power when triggered by surges in the dimensional fields. One such is under the Sargasso Sea in the area you refer to as the Bermuda Triangle ... not in the waters per se, rather below the sea in a remnant of the tunnel system. These exist in a dimensional lock and can only become physically visible in 3-D at certain times of excessive magnetic activity.

The catastrophe occurred when six of the expedition members were unfortunately exposed to severe radio magnetics that essentially dematerialized the two nearest the unit when the generator auto powered on due to a tectonic energy surge. The other four were critically damaged from the residual effects of the tremendous dimensional warping fields produced. The effects were similar to the injuries sustained by those in what you term the Philadelphia experiment. Kinkaid and the academics died within hours, despite hopeless efforts to save them. Only the two military men posted at the surface entry survived. Both were deeply frightened. (Subsequently, the relay complex was put into a frequency that effectively placed it in a dimensional lock.)

Afterward, the government sent a third and fourth party to the entry area and sealed it. The area is still off limits. In the early 1950s the area was opened, reentered and explored by government agencies.

Question to Metatron: You mention the 'Atlantean Tunnel System,' did the Atlanteans have access to the Grand Canyon? And if so, for what purpose?

AA Metatron: *Indeed! There was access not just to the Grand Canyon, but virtually to every area of the planet, both surface and interior. But while Atlanteans utilized the system, it was not of Atlantean construct. Many of those you term Atlanteans were not of terrestrial origin. It should be understood that the majority of those involved in the technological development of the tunnel system were of Sirian, Arcturian and Pleiadean origin, but they lived in the Atlantean areas and interacted quite freely and openly with those of earthen biology. Many of you, including the channel, experienced Atlantis originally as extra-terrestrials. Eventually many of these took on human biology ... becoming entrapped somewhat within the time sequence of your Blue Planet, becoming captive to the grid in essence.*

The Grand Canyon was recognized initially by the Atlanteans for its myriad mineral deposits. Of prime concern was an invaluable phosphoric compound infused with akash. This compound was very rare, only occurring in certain conditions and formations. The Atlanteans had the technology to locate, congeal and preserve this living semi liquid akashic compound, blue green in hue. In the vernacular of the ancients, it was termed as the philosopher's stone. It had a relatively brief 'shelf life,' and once preserved it was rapidly transported to the Temple of Healing in Poseida. It was an essential ingredient in the complex rejuvenation process for healing terminal maladies and for vast expansion of mental capacities. It was highly prized and is mined yet today by those of the Pleiades, Sirius and as well as mankind from the future. Does that surprise you? This scarce compound exists primarily in the areas of the Grand Canyon, Arkansas, Brazil, Tibet and eastern Siberia. Its greatest abundance remains in the area of the Grand Canyon, where its matrix compound was seeded. It's greatly prized value was the initial reason for the extensive tunneling to the area.

In time, all of the attributes of the Canyon were recognized and utilized. The Atlanteans as well as the Lemurians had communities in the Canyon chasms. Although most were initially for mining, others were developed soon afterward that included temples, rejuvenation facilities and what you would term monasteries for the guild of scientist priests. The tunnel system was completed approximately 38,000 years ago, in your current measurement of time sequence, soon after the initial construction of the Great Pyramid of Giza. The pyramid observed in the Canyon was part of the Temple and Monastery complex.

There were and are many hyper dimensional generating units, like the one that Kincaid's party discovered, in the area. These units are and were of Sirian technology. Some were for transport to other relays within the tunnel system. Transport relay stations were dispersed along grid points to certain locations such as

Egypt, Tibet, Poseida and Og. Some of these were specifically for material movement, and other prototypes were for transport of personnel through a hyper dimensional warping within a 4th dimensional matrix. Others were developed for movement to and from the future, used essentially for time travel, in your terms. We have told you before that aspects of Atlantis were truly from the future, pulled into the past.

The Atlanteans calculated, designed and constructed earthen grid works within architecturally created sacred geometric grids, utilizing natural telluric powernodes to assist in stabilizing and maximizing the energy flow within these symmetrical grids. These are in effect earthen grid systems somewhat of the 'Reshel' classification. One of these connects the Canyon's Temple of Isis to the Pyramid at Giza. An energetic resonance exchange occurs here and in other areas of the greater vicinity within these numerous earthen grids. Boynton Canyon near Sedona is another of these that is very 'Egyptian' in visual feel and resonance. Within other sections, the axial tonal and Reshel type grid connections are to the Yaluzhangbu Grand Canyon located in Tibet. That is why so many of the formations in the Canyon were given Egyptian and Buddhist names, albeit in modern times by explorers who were consciously unaware of the reason they felt compelled to assign such appropriate nomenclatures to the formations. The resonance and connection to these places is quite real, very tangible. In others areas of the greater Canyon vortex, the axial tonal grid connections are to the Lake Titicaca area of ancient Og, modern day Peru and Bolivia. The underground chasms and tunnel network are equally vast in the area of Cusco and Tiajuanaco. Although units for transportation were available, certain adepts were able to translocate within these fields between the locations mentioned, simply by utilizing the Reshel style grid system. Indeed there are several potent pyramids built by the Atlanteans within the Canyon; although most are now covered by the sands of time, one yet remains on the surface.

There are certain groups today whose purpose is to rediscover the design and utility of Reshel grids. They are well along this path, yet much remains to be understood. Reshel grids are in essence a stabilized, self generating energy pattern, utilizing specific geometries. These not only link powernodes but also are the frequencial keys to the planetary and stellar grid system. These are capable of housing stargates and time gates.

The tunnel network of the Atlantean past was in its prime for over 25,000 years. It fell into disarray after the demise of Atlantis some 12,000 years ago. Subsequent tectonic movement and shift collapsed the global network, although some portions are still intact. Few are maintained. The Sirians and Pleiadeans in specific areas have reestablished some of these. The inner realm Lemurian races no longer need them, having ascended to a state where they are capable of translocation through pure thought frequency.

We will add that the inner world chasms that exist below the canyon are among those closest to the surface of the planet. These extend for literally hundreds of miles and are not only occupied by the blue humanoid race of highly evolved Lemurians but also contain bases for those from the Pleiades, Sirius, Andromeda and Arcturius. One of these connects to a massive chamber below what you term Groom Lake in Nevada. This area will become more important in the millennia to come, as it will serve as an evacuation point when the necessity arrives to seed other planets. There are interactions between the four primary extraterrestrial groups working within the benevolent Galactic Federation with scientists of certain governments today in this area. The work with the Federation is for the benefit of mankind.

As we mentioned earlier, in millennia to come this area will be key in the technical development of the seeding of enhanced DNA humankind into other worlds.

Preserving the Canyon

Now, during past periods the river course in the canyon was much longer, double its current length. It was shortened considerably over time through volcanics and tectonic shifts. More recently in your times the canyon was flooded into lakes by the placement of reservoirs at Lake Powell and Lake Meade with the construction of the two dams. The area below what is now Lake Powell regrettably was among the most stunningly beautiful sections of the Grand Canyon and contained energy fields that are not replicated in the same manner elsewhere. However, the green copper waters held within Lake Powell remain quite bountiful in healing quality, and all who recreate in this lake benefit immensely.

There will be a time in the not too distant future when water shortages will motivate corporate government concerns to build a third dam. This must not be allowed to happen.

Be Well and Trust

We close by telling you that the Canyon should be viewed as a great benevolent energy, one that enhances all who visit. There are many such places on the Earth, but none quite so varied, quite so concentrated in auspicious advantage. Dear Ones, your planet is on course, it is well, it is where it should be. So often you speak of healing the Earth, yet we tell you it does not require healing, rather the Living Planet is healing you. The changes that are coming do not require healing, they are not the result of a planet in disarray. Gaia is strong and on track. The planetary transitions are in truth a graduation required in order to hold the higher dimensional matrix needed for the Ascension. These transitions must happen, but they need not come about in global cataclysm as occured at the end of the 3rd world, in the sad demise of Atlantis. We tell you the current path is that catastrophic change will not occur on a global level. There will be regional events, some you have already seen. The polar shift has already begun, and it can complete itself in a relatively smooth transition.. The polar caps will melt, the water will rise, but this must take place in the new paradigm. It is the rebooting. Trust this truth. Be well, be happy, and find joy in these rapidly ascending times. Treat each other with love and respect! Honor SELF! You have waited a long time to be here to guide this heralded Ascension into the 5th world and 5th dimension!

Many areas on your planet are already in the 5th dimension and beyond. Many of these infinity points offer you profound insight, renewal through love and joy. The Grand Canyon is among the most potent of these. Indeed it is the generator, the wind beneath the wings of so many powernodes in this sacred region, which include Zion National Park, San Francisco Peaks and Sedona. Truly the Grand Canyon is the Womb of Gaia.

I am Metatron, and you are Beloved!

… and so it is. AA Metatron via James Tyberonn

Sacred Redpath ~ The Eagle and the Condor

Archangel Metatron Channel via James Tyberonn

And as Creator spoke of understanding, I looked up and saw the rainbow leap with flames of many colors over me. And I saw the sacred hoop of my people was but one of the many hoops that made one great circle, wide as daylight and long as starlight, and in the center grew one mighty flowering tree to shelter all the children of one mother and one father. And while I stood there I saw more than I can tell and I understood more than I saw; for I was seeing in a sacred manner the shapes of all things in the spirit, and the shape of all shapes as they must live together like one being ..."

BLACK ELK, Wichasha Wakan
(Holy Man) of the Oglala Lakota

Greetings! I am Archangel Metatron, Lord of Light, and I greet you! We welcome you here at this moment in your space and time, and we send each of your beingness a ray of supreme love.

JT: Greetings, Metatron. While I have been involved in metaphysics most of my adult life, it was actually the Redpath, the series of four annual five day vision quests and shamanic journey ceremonies that seemed to accelerate my spiritual experience. Can you speak of these modalities?

AA Metatron: *Indeed! Although you have had many lifetimes in the sojourn of the Native American, it is a greater thread that draws you and others to this great society. Truly of all the realms existing throughout the history of the planet, the golden era of the Native American was one that despite* *hardships offered opportunities for prodigious spirituality, wisdom and inner peace. It was indeed a brief but golden era that many many are still drawn toward, you see. It afforded mastery to those that sought it in a more feasible manner than any other society or culture has ever achieved since the long passed golden age of Atlantis. Many great souls and avatars were part of this epoch. It was brilliantly unique in the manner of incorporating the beauty and consciousness of what we would term the natural world and all its kingdoms.*

Prayer Fast and Vision Quest

Specifically that which is termed the prayer fast ritual has been quite prolific in exponentially advancing many seekers to gain access to the inner worlds and higher dimensions of the God mind. We will say that many Earth-Keepers first gained such knowledge through the uniquely effective modality of fasting termed the vision quest while embodied in the Native American lineage. These were lifetimes of great advancement. And so this is true for many Lightworkers and Earth-Keepers now on the planet involved in the work of the Ascension.

The vision quest is a sacred and ancient technique of prayer fasting. It evolved into its four year ceremonial practice with the Native Americans, but in fact prayer fasting has been a means engaged since ancient days by all humankind to seek the inner God. The Lemurians, Atlanteans, Egyptians, Hindu and Buddhist were among those that rigorously utilized supervised disciplines that involved fasting. It occurs in almost all of your mystical sects even today for devout seekers.

The indigenous of the red race evolved and adapted the vision quest as an ancestral multilayered mechanism for teaching seekers to become luminous beings. In addition they found certain sacred plants could be ingested as sacraments to effectively accelerate and enhance access into the inner world. Many of you, including the channel, used these methods in many past sojourns. So we tell you that these modalities were of great benefit when conducted in sacred intent. Fasting in natural solitude is a practice that clears the frontal mind and opens and punctuates that which you term the subconscious. Vision quests involve clarity of intent with meticulous preparation, and all the sacred kingdoms of the celestial and earth nature became the teachers. Indeed, it is a great travesty that much of this sacred wisdom has perished in the industrialization of your world.

Plant Teachers

Question: You mention plant medicines as teachers; were plant medicines a part of that accelerated growth?

AA Metatron: *For some, indeed for you and many of the Native American culture, it was exactly that, yes. Their profound campestral knowledge of the plant kingdom made it so, you see. However plant medicines, we add, are of benefit to some and not of benefit to others. To some they can enhance and amplify great inner journey, to others they can be septic. There is no plant within the plant kingdom of the medicinal genre that is of benefit to all. It has always been the nature of what is termed plant teachers and plant healers. To some extent this will vary according to the corporal consciousness as well as the system of belief of the individual. It is for that reason that the spirit of the plant itself was acknowledged by the shamanic elder. Intuitive communication with the spirit of the plant was essential in order to know which seeker would react well, which would be toxic and which would obtain no interaction. Every member of the plant kingdom has multiple dimensions of its awareness and its purpose. Each relates in a variety of ways and purpose* *to its environment and to humankind. So to those who sought to define their purpose, in narrowing or expanding the conscious focus, certain psychoactive plants made themselves available for multidimensional lenses of viewing the sojourn.*

Just as the ancients knew that certain aspects of the mineral kingdom offered multidimensional experience, so were they shown that likewise the plant kingdom offered itself for greater journey. Those tribes that lived in the forests, plains, mountains and deserts fostered shamanic masters who had the ability to accurately intuit knowledge of plants, their medicinal values and boundaries and to work with them as teachers and healers. Others acquired mastery through communications with the consciousness of the directions, elements and animal kingdoms. Some, termed shape shifters, acquired the ability through disciplines and ceremony to meld their etheric consciousness within that of the animal kingdom. So then these shaman became specialized in the abilities, some became the savant vessels of the plant kingdom, and much was shown to them. They became aware of the properties of the leaves, the roots, the flowers, the barks, understanding which to blend and how to use them.

Earth Tribe Wisdom

The system of spiritual advancement and wisdom you term the Redpath was truly not just to the red skinned races, it was developed by all races that lived bucolically and tribally in nature. But we will say that this era of the Native American was a culture that excelled in many ways. In the Americas the indigenous tribes developed a great knowingness of what areas of the Earth afforded higher dimensional energy and what plant could assist in that process. So the practice of the vision quest, of the purification lodge, of the plant journey were used to teach, just as the animal kingdom and mineral kingdom were also recognized as being great teachers and providers. This was the gift of the pastoral life. Sadly, it is a knowledge that few now hold.

The Eagle and the Condor

Question to Metatron: Can you speak about the prophecy called the Eagle and the Condor?

AA Metatron: *This powerful prophecy was sent from an energy entourage of what is termed Star Nation. It was given in varying tiers to the indigenous beings descended from the Lemurian red race of the western portion of the Americas, including those whom you call the Hopi. In this prophecy of the Eagle and the Condor, the Eagle represents the energies of the North and the Condor the energies of Latin America. Both were once as one, and the prophecy states that when the Eagle and the Condor again fly together, it will signify the opportunity for the return of wisdom and balance.*

So the prophecy speaks of the current time in which the two lands will mix in a benevolent contract, sharing their messages and ceremonies once again with all the people in order to gather in renewed strength and reactivate that which was lost, that which was sacred. You see, it is the sacred wisdom of the Condor, the indigenous of Latin America, that has not been as scattered, we might say. The ancient wisdom of the Condor must go north and that of the Eagle south. It is to heal the Eagle. That is why those Elders, Earth-Keepers and Lightworkers of the North are drawn to ancient Og, the Land of the Sacred Lake now called Peru and Bolivia, and those of the Condor are drawn to the North at this time.

We tell you that in the most ancient days it was the Condor that represented the north and the Eagle that represented the south. Both winged beings are inter dimensional and were revered as great wise beings and divine messengers. Indeed these are among the most noble of avions. The prophecy speaks of the united peoples opening their hearts and hands to erase oppression, exploitation and injustice. The avions will etch the sacred infinity symbol of the number 8 across the skies of heaven to signal the symbol of renewed freedom. The union of the Condor and Eagle is occurring now, as we come to the heralded Ascension of 2012. Now it is the Eagle for the north and the Condor for the south. When they turn to fly in retro union, the symbol of infinity will cover the hemisphere of the Americas, you see?

We offer you another simple yet complex addendum to the prophecy, "Without the Condor, the Eagle will die."

For the Eagle to fly, much must be done by Earth-Keepers and Lightworkers in the United States. The Harmonic must be raised. The United States has many powerful spiritual pockets, and these are growing in inertia. But, the great masses of the people in the United States are incased within gilded cages of manipulated misunderstanding, victims of their own insulation and macho narcissism. Few Americans have sovereignty. The common man in the United States is enslaved by debt, manipulated by media and intoxicated by quantity. Sadly, America has become a corporation based on profit. The people of America are no longer citizens, they are employees. Many are beginning to wake up and are working to change this, but it will not be easy.

Now, because the governments of South America became colonized into Castilian aristocracies, a great separation polarized between the controlling patricians of the rich and the stark poverty of the indigenous. Yet despite this inequity, and partially because of it, a larger degree of the indigenous peoples retained their campestral culture and knowledge ... especially in the sacred land of Og, where the indigenous compose a larger segment of the populace and to a smaller degree in the lands of the Yucatan. In South America, indeed in this time of the fulfillment of the Eagle and Condor, much transition is now afoot. There will be a time in the not so distant future in which South America will displace North America as the primary bread basket of the world. As its economy strengthens, so will emerge a powerful cry from the common man for a stronger voice and more equity among all its classes and people. This is beginning now in Bolivia, Colombia, Peru and Venezuela and has been evolving in Argentina, Brazil and Chile for more than a decade as a growing middle economic class emerges. The Condor of the mighty Andes is strengthening!

While Og was originally Lemurian in populace and settlement, it became more Atlantean after the cataclysms. Many from Atlantis migrated to this area as well as to parts of Brazil as the lands began to shake. Just as colonies of Og such as Tiajuanaco still remain in present day Bolivia, so Atlantis had many colonies in Brazil. The largest were in the areas of the massive quartz and mineral fields. There are yet today remains of Atlantean ruins in Brazil in Ceara, Rio Grande do Norte, Piaui, Mata Grosso, Amazonas, Goias and Maranhão. Some are yet to be found in Bahia and Mineas Gerais. The area called Sete Cidades indeed housed a mighty station of the crystal power system. The Atlantean tunnel system still connects Og and Brazil, which in ancient days was referred to as the Lands of Muir, Mirubi in the north and Il Barr in the mid section.

As we have told you, the beauty and strength of the indigenous cultures lay in their understanding of the conscious nature of the earth, the elements, the directions and the unity of all things. Some have retained such knowledge.

There are indeed those in Latin America, primarily of the Starseed, the indigenous wisdom and medicine keepers, who are assisting in the healing and awakening experience. Certain sacred plant ceremonies and shamanic journey quests that allow for multi dimensional clearing and understanding the sovereign divinity of self are still available in the land of the Condor and no longer in the land of the Eagle. The culture of facilitating these sacred plant teachers has essentially been removed in the United States.

As these cultures were disintegrated by the European settlers in both North and South America, much was lost especially in the land of the Eagle. However, there are now great elders called forth to help restore these lost elements of wisdom and to reactivate this knowledge in the rainbow tribe. It is not a paradox that some of the great wisdom keepers and elders from the golden era of the red tribe are now reincarnated in other races. Such is the nature of reincarnational cycle, and indeed the reemergence of these elders is part of the reactivation phase required for the Eagle to fly.

Few of the indigenous cultures had written languages, so much of the learning was passed in oral tradition. As the tribes became conquered in war, they were scattered and forcibly immersed into European and colonial teachings. Much wisdom was diluted and forfeited, as millions were indoctrinated into Christianity and absorbed into industrial culture.

Starseed and Star Nation

Although the indigenous societies did have a great understanding of the living Earth, their wisdom schools existed in small enclaves. Purification lodges, vision quests and medicine journeys were conducted by highly evolved souls. These elder shamans had attained the ability to increase their frequencies for merkabic travel into higher planes. In this merkabic resonance, they achieved the ability to etherically navigate in timeless multi-dimensionality. Much of this was guided by an etheric entourage of Starseed, Ascended Masters of Sirius B, that became known to the indigenous as Star Nation. This discipline was earned. It was not taught to the populace. Enlightenment has never been achieved in aggregate, rather it is for the select few who have persevered and chosen to work devotedly for such Mastery.

A great guild of Ascended Masters reincarnated in a cycle from Mu and Atlantis within the matrix of the indigenous tribes. These omnipotent beings became known as the Clan of White Eagle in the north and the Clan of the Golden Sun in the south. Both cultures achieved much through many centuries and provided templates for great learnings to the true seekers in the Americas. This was assisted by holograms of coded light programmed into golden discs that were transplanted to the areas of Og, Yucatan, Mongolia and Sedona after the demise of Lemuria. There are in fact 12 golden discs globally placed by the Star Nation of Sirius B, and these are distributed across the planet.

However, in the dissolution of Mu some of the discs were transplanted. In the subsequent fall of Atlantis others fell into dormancy, and the harmonic was further reduced. All were affected. As such, the people became somewhat lost for a period and dispersed into segments. Their systems of belief evolved differently within ritualized ceremony and tradition.

The Breakup of MU and Atlantis

Now, it has become quite fashionable to view the culture of the Lemurians as being quite highly evolved, and we tell you that while the Lemurians did have a splendid phase of great wisdom, it was comparatively short lived. The arcadian society of Mu did not in fact reach the heights of utopian spirituality that are currently credited to them. In truth the Atlanteans of the Law of One, in the Golden Age of Atlantis, reached and sustained far greater spiritual heights than that of Mu.

The societies of the indigenous cultures of North and South America likewise reached heights of wisdom for a brief wondrous era. Great connection existed between the lands of the Eagle and Condor and continued for a while even after the breakup of Mu. Those of Central America and Og had considerable interaction with benevolent extraterrestrials, as did those of Atlantis, and achieved a phase of vast astronomical and scientific knowledge.

But as the massive tsunamis of Atlantis swept over the lands of Og and Yucatan, much was destroyed. A gradual downward spiral began all over the planet. Factions of the survivors in Og and Yucatan etherically ascended, others went into the inner chasms of the Earth. Those that remained evolved different languages, and essentially the unity of all tribes was disconnected in the passing millennia.

So, long before the advent of the European invasions, the people became fragmented. Basic survival became the tantamount priority. Pockets of groups including the Hopi, Toltec, Mayan, Navajo and Lakota retained etheric interaction with Star Nation and as such their high spiritual practices and ideals. Many others did not. In time, hardships brought tribal wars, cruel tortures, and some denigrated into the corrupt practice of human sacrifice. Indeed, this was not just in the Americas, it was global. Much was lost in time.

So we tell you that potent and brilliant gift of the red path, and indeed the hope of the Eagle and Condor prophecy, is the reactivation of the ancient schools of disciplines offered to the individual and of the knowledge of the living planet.

The Rainbow Tribe of All Mankind

What is presently misunderstood by many is the reincarnational cycle in the progression of man's sojourn. This is not to point fingers, because most of your religions either dismiss or misunderstand mankind's reincarnational cycle within multidimensionality. If this concept were better comprehended, humans would perhaps truly understand the importance of all tribes as the unified rainbow tribe of all mankind, of every race. You see, while physical tribal heritage should be honored, the soul's reincarnational heritage is truly of much greater bearing. Truly all of you have been every race and every gender.

So when ancestry is honored, know that it is an indispensable truth that some of the extraordinary elders and medicine keepers of the Native American past are now walking the Earth in bodies of all races. That is why so many non indigenous people are drawn to the wisdom of the Earth knowledge and wisdom disciplines of the Redpath! You see, it is because they have lived it in past lives and still live it in the NOW of multidimensionality.

In kind, those that understand the law of balance of the multidimensional Self will understand that many of the conquistadors, officials and gold miners that took away indigenous lands are now in red bodies in order to balance those very actions. There is beauty in this truth!

A great debt is owed to those souls who have chosen indigenous lifetimes, both past and present, in order to pass down and maintain these sacred disciplines. Then and now, these lifetimes have been ones of devotion and hardship, choosing to live on the land amidst prejudice and economic hardship, often away from the trappings and luxuries of wealth.

So the Eagle and Condor prophesy is to involve all races of the tribe of mankind in reactivating what was great in ceremony and teaching. Indeed there are great elders in both red and rainbow skins that have been called by soul contract to unite all peoples as part of the fulfillment of this prophecy. This task is daunting, and much respect is owed to those souls who diligently work to achieve this balancing unity. For the rainbow tribe of all mankind to drum, sweat, dance, council, quest and sing the unification songs of prayer, honoring the sacred lands, elements and directions in devoted reverence is of benefit to mankind and to the Earth. Do you see?

Spirit, Religion and Dogma

The channel asks about dogma and protocol. Dear Ones, we tell you that no one religion on your planet carries all the answers. Truly, spirit is found in the church of the heart through the intent and impeccability of the individual. Spirit knows no dogma, group consciousness fields and systems of belief form dogma. The paradigm is love, and love begets unity. Love yourself, and be kind to one another, for disciplines of enlightenment are as varied as the flowers of the field.

There are many paths for a man to walk, and all those of integrity will find their light. So it is always true that discernment is required.

Disciplines in the wisdom schools of ancient ceremony evolved in time. Indeed systems of belief in some areas are being refined somewhat in the new paradigm. Integrity of intent is the key. Spirit acknowledges all! Indeed you are all sparks of the divine!

Masters, the flesh offering tradition which has been utilized by many religions including Christianity, Islam, Redpath and Hindu is a choice based on tradition and the group consciousness of belief systems. There was a time in which such disciplines of mind control over body physical had its seeming logic, for the winters were harsh, and the summers were dry. Food was scarce. Developing the will to sustain such hardships in ceremony served to allow the body physical to be controlled by the mind, you see? But there is a fine line of important separation between 'flesh offerings' that are intended to control body through mind and self flagellation based on the erroneous belief of human unworthiness and original sin, you see? Intent is key. So it is an individual choice of free will in duality based on belief, and therefore it is not to be judged. But we tell you in the new paradigm it is truly neither appropriate nor required.

But the restriction of the female, exclusion or limitation by race and fear of reprisals are inappropriate in any dogma. Beloved, this applies in all your religions. Elders of the rainbow tribe of humanity on all true paths are revealed by the love and integrity of their light quotient. You are all Elders of your Divine Nature. You are all sovereign beings, and we honor you! This is the rainbow of the Eagle and Condor. You are all sparks of God. Spirit sees only the heart. So we honor all of you in those disciplines of your choice in developing the inner eye, of communicating with the living elements, directions, celestial beings and in developing unconditional love and understanding of the unity of all life!

The Unity of Titicaca and Sedona

Now, for the Eagle to fly, an essential part of the heralded Eagle and Condor reunion involves the hemispherical planetary balance. Certain key powernodes in the land of the Eagle, North

America, are clouded and are being revitalized from the land of the Condor. The key powerpoints that are aligned to the United States from South America are assisting in this vital balance. A great connection has always existed between the vortex-portals of Sedona, Arizona and Lake Titicaca. You see, the two are meant to flow in symmetrical counter balance. But that ancient energetic ley route has become less effective, and it is now to be repaved and retread as 2012 nears. That portal which is called Lake Titicaca is much more intact and more capable of feeding and healing the other.

Some of the once potent centers in the North are diluted, surrounded by commercial interest. As such, clearings and revitalizations need to occur, and many new portals are emerging. Do you recall the biblical allegory of Jesus casting out the merchants from the Temple? You see, portals and celestial points are living Cathedrals because of their natural magnetics and alignments, and they were never meant to be areas of commerce and housing. That is why indigenous caretakers designated certain sites as Hallowed Earth, and only adepts were able to live within the energies on a permanent basis. That is why many of the potent vortex-portals in the United States, such as Monument Valley and Grand Canyon, were designated to the care of certain indigenous tribes. In later years others were protected through divine guidance of inspired human beings into reserves you call National and State Parks. Indeed this is part of the Eagle and Condor. It is why many from the land of Og are drawn to the Eagle, and many of the Eagle are making pilgrimages to Lake Titicaca.

Lake Titicaca is among the most pure of portals; it is a potent battery for both receival and transmission of coded higher dimensional energies required for the Ascension of both the Americas and indeed of the planet. Lake Titicaca corresponds to the Ascension in many different ways. It is a consciousness that enlivens each dimension and relates to the individual within a specified frequencial resonance of these dimensions. Do you understand?

At Lake Titicaca these frequencies are fixed to humanity and more readily available to those who align with them. Sedona is being directly boosted by an axial tonal line from Titicaca. You see, Sedona and Lake Titicaca are fixed to certain beings among humankind and to the very consciousness that directs the planet. Both are very potent and very important to the Earth grid, the celestial heavens and indeed to mankind.

And so the axial tonal line connecting Sedona to Titicaca will remain connected as an umbilical feed of sorts until such time that an upshift will complete itself. Just as humanity will remain in the 3rd dimension for a time until the shift comes, so will Sedona be brought into greater energy in the Ascension.

We will say that the energetic field, the very length, breadth and vitality of the vortex of Lake Titicaca is one of the largest on your planet. It is important that you understand that Lake Titicaca is energetically balanced; it is neither male nor female, it is both. That is why many are drawn there. We assure you, if it was either one or the other, the feminine energy or masculine, Lake Titicaca would not draw that amount of power, it would not contain the equilibrium, you see? It contains proportional levels of both, and as such the potency, the harmonic and the harmony.

So the consciousness of this vortex will balance any human that is in imbalance. One who is overly macho or overly feminine will receive the opposite to obtain the balance. In your current scientific understanding, you might think that the presence of anionic and cationic, female and male, would neutralize one another in bonding. It would seem so, but that is not the case. Rather, they activate one another. This rare point on the planet which is BOTH a human chakric center and planetary chakra consciously directs and regulates itself according to what is needed for balance, do you understand?

And so the Condor of Titicaca feeds that harmonic equivalence into the Eagle of Sedona ... and in time both will fly! And in so doing, they will awaken the golden peace that many of you found so fulfilling in the brilliant and compassionate heart of the Redpath in this dawn of Earth's long heralded Ascension. For truly now is the time of the Rainbow Tribe of All Mankind to stand in unity!

In closing we say again, please know how deeply you are loved. We ask you to walk in honor, respect one another and allow each the freedom to find the wondrous light of Divinity that shines in each of your hearts.

I am Metatron, and you are Beloved!

… and so it is. AA Metatron via James Tyberonn

"Oh, Great Spirit Wakan-Taka
Whose voice I hear in the winds,
And whose breath gives life to all the world,
hear me, I come in humility,
I ask for your strength and wisdom.
Let me walk in beauty and make my eyes ever behold
the red and purple sunset.
Make my hands respect the things you have
made and my ears sharp to hear your voice.
Make me wise so that I may understand the things
you have taught my people.

Let me love all mankind as my relatives
Let me learn the lessons you have
hidden in every leaf and rock.

I seek strength, not to be greater than my brother,
but to fight my greatest enemy - myself.
Make me always ready to come to you
with clean hands and straight eyes.
So when life fades, as the fading sunset,
my Spirit may come to you without shame."

Lakota Sioux Chief Yellow Lark

Envoys of Light ~ Back from the Future

Archangel Metatron Channel via James Tyberonn

Greetings, I am Metatron, and I welcome you. I welcome you with words of salutation and with an energy that encircles you, an energy of warmth and love, an energy of what you are, have been and will become. I greet you in the eternal moment of NOW!

Now, many of you involved in the Planetary Ascension find yourselves in a phase of Earth time that you cannot fully fathom, cannot totally embrace. It is at times as if you are a visitor here, far from home, and your energies do not exactly fit into the Earth of 2008 and beyond.

Dear Ones, know that these feelings are valid and are in a true sense culture shock, as you awaken into the multidimensionality that is unfolding before you. You see, many of you, in your terms, are indeed from the future! There are a vast number of souls who have chosen to come to the Earth at this time from dimensions beyond the fifth from both the future of Earth and from the future of higher dimensional worlds, that will become better understood to you in linear time progression.

You are indeed Envoys of Crystalline Light, Ambassadors of Love. You have moved back from the future, in your terms, in linear Earth time to assist mankind in navigating through the tumultuous times that have unfolded in the last forty years and the times that will be presented in the next twenty. You see, it is an indelible fact that you exist within multidimensionality in all lifetimes simultaneously. As such all of your 'past' and 'future' lifetimes are occurring NOW in the higher aspect above linear time. And many of you more advanced souls have in essence returned from the future to help assure the Ascension.

You began accumulating in mass in what is termed the 'Baby Boomer' generation, the post WW II era; and your numbers have steadily increased into what is termed Generation X, Indigo and Crystalline children. You are the Envoys of Crystalline Light, the pioneers of the alternate Earth ... the Ascension. You have had many roles, many assignments and many duties. You are the Wisdom Keepers, the Earth-Keepers, the Shamangelics, the Fire Dancers, Dream Walkers, Warriors of Light, Shape Shifters, Star Beings, Elders, Strength Bearers and Healers of Time!

You came here to engineer the reconstruction of a different outcome here on the planet of duality, the Earth! You are here to guide mankind through the parallels of probable calamities and potential catastrophes that otherwise could wreak havoc, as has happened so many times before through troubled waters. You are here to make sure history is NOT repeated. And we tell you that you are succeeding.

You see, you are the troubleshooters, the renegades, the paradigm busters! We use the analogy of the sea to clarify this concept of the Envoys of Light. Imagine the Ascension as a sea going vessel. The Light Envoys are expert navigators who know exactly where the currents are perilous, where the tides are crashing into rocks ... and the Envoys know how to avoid them, where to sail past, how to maintain ballast. They are the teachers with expertise in many fields, and in your vast gathering you bring a collective wisdom that is unparalleled on the Earth at any time in the planet's history.

You see, many of you have experienced the Ascension of other planets and other worlds, several times; and you know the perils. You understand the mechanics.

We tell you that the Earth of 1970, 1980, 1990, 2000 and 2008 has made great progress; much has been diverted, much is on path. The Envoys of Light have made a difference, but much remains to be done.

However, you now see a great deal of events around you that are confounding and quite confusing. You look at the condition of mankind, and you know it is a far cry from the Futures of which you came. You are not comfortable with where you are; it does not quite feel like home. You see wars, poverty, corruption, pollution and hopelessness around the globe and you ask, "How have we made a difference? How can this planet be Ascending?"

You see a world controlled by power and greed and the masses of society in cages of manipulated ignorance and fear. You see nations blinded, and the majority of people intoxicated by quantity.

Is it any wonder you don't feel 'at home.' As Envoys of Light you know deep within that a better society can and will exist, because you have been there. It is in the future Dear Ones, and you came back to manifest it. Don't doubt it! Don't lose hope!

Dear Ones, listen carefully, and have faith in the truth we share herewith: The Light Quotient on the Earth is at a higher level now than at any other time in the planet's linear history.

Does that surprise you? We tell you that approximately ten percent of the planet's population of some 6.7 billion souls now carries sufficient light to manifest the Ascension. We tell you that you are not only on schedule, you are slightly ahead! The work you have done and will do, Dear Ones, is right on track. You are well beyond the halfway point. You are well on your way to achieving what you came here to do!

Now, much remains to be done. There are still rough waters ahead that require expert navigation.

The great majority of the Envoys of Crystalline Light have been drawn to the metaphysical world from all walks of life. Many have chosen to focus on spiritual growth and have selected humble paths, choosing not to be involved in what may be termed the material aspects, choosing to be energy holders of the Crystalline Light that powers the planetary graduation that lies before you. Know that we commend you, all of you, for this noble and necessary task.

But we tell you, there are indeed many Envoys of Light that have chosen to make changes from the inside of what you may term the 'traditional establishment.' Many Envoys of Light, many advanced souls from the 'future' are now in positions of leadership in governments, in corporations, in universities, in the sciences, in the arts and in medicine. But these are not leaders in the traditional sense of the status quo. These are not paradigm busters from the school of shake, rattle and roll! For they know such a path would limit their ability to achieve the change from within, you see.

You will not often find these individuals drawing followers around them to protest the current state of affairs in open rallies to incite discontent. Rather it is their wisdom and nature to seek change through achieving credentials, achieving grounded wellness and then projecting their influence through the example of a life of confidence and balance within the 'establishment.' These Envoys inspire wellness through living it; they gain respect through quiet demeanor and example. Accordingly, others are drawn to seek their council and hear their opinions.

These are wisdom keepers with the ability to see issues from many perspectives and discuss changes and shifts from a broad stance. Truly these are teachers from the grandest sense, because their thoughts are expansive, and their experience has been varied and prolific. As Envoys from the future, they have little to actually learn from the current sojourn but a great deal to offer in a purposely chosen demeanor, that of seeking change from inside the complex organizations of the governments, academia and corporate worlds. Truly a delicate task, for within these power structures are many of great wealth who would seek to block them.

Now, indeed there are Light Envoys who have chosen a more pronounced voice. These are brave souls of crisp clear energy and are quite easily discerned: many from that termed the Arts and Entertainment, some in medicine, some in political arenas. Quite often these are the energizers that tend to mock the status quo and demand change. They draw many behind their leadership through their high public profile and achievements, yet often are they termed as radicals and essentially become easier targets for those that would block them. Do you see how it takes both the front line and back line to truly teach change?

You must understand, Dear Ones, that the 90 percent of the populace who do not carry the bright Light Quotient are just as valid in their soul aspect as are you, the Envoys. The masses of the status quo will not be changed through bulldozing: they will not be changed through force! Rather they will be changed though Light, through Love. Yet both approaches of Light, front and back, serve to the whole.

An example of this can be derived from the life of the great soul called Martin Luther King and his counterpoint termed Malcolm X. One sought peaceful change, the other change through violent revolution. In a valid sense, they were opposite sides of the same movement. In truth, Malcolm X, through his preaching of extreme methods caused many more to see the greater functionality and greater truth of the nonviolent methodology taught by Martin Luther King. Without Malcolm X, Martin Luther King would not have achieved as many followers, you see.

The movement for racial equality that began in the 1960s in the United States is quite an example of how different advanced souls played remarkably varied roles in a concerted group effort to open the eyes and hearts of the majority masses. A great number of brilliant souls, very advanced in the art of sound, reincarnated in ethnicity during this period to create a great wave of popular music you termed Jazz, Rhythm and Blues and Motown. The contagious artistry swept the world and was embraced by all. Even those of the prejudicial culture were forced to quietly ask themselves, how could such musical brilliance, music loved by all, be created by a 'lesser race.' Thus the wheels of change began to gradually turn in the most varied and subtle of mechanisms, for truly all of you live lifetimes in all races, all genders.

The point we wish to make is that the changes sought on the planet are being worked from inside and outside the status quo. There are great souls working in tandem, in concerted synergy from many sides. Change will come, but ultimately true change will come through light, through the growth conceived by love, not through force.

So whether you are an Envoy in a modest life of quiet spirituality or an Envoy of high profile, both have very valid and very potent purpose. All of your efforts exponentially add to the Quotient of Light. Remember, the ten percent has changed the course of the planet. Love is a far greater power than hate or greed. Consciousness is far greater than ignorance.

So we tell you, do not just dream the 'impossible,' you must also envision it! That is why you are here: hold that dream, carry that vision daily and exude the LIGHT! Not through forcing your will on another but by shining your Love to ALL! Perhaps when you are asked about your opinions, some will shake their heads and say, "This one is totally delusional, completely out of touch with the real world!" But Dear Ones, as Envoys from the Future, how could it be otherwise? Your hearts are aligned to the future, and the future is indeed a beautiful dream: a dream that comes true, a vision that is being birthed into manifestation through the hearts of the ten percent!

You perceive a world in which mankind cares for one another, a world where all are equal, where all are cared for and all are given opportunity to grow and love and live life in joyous unconditional LOVE in the dimension of five and beyond. Truly, Unconditional Love cannot exist in the dimension of three. It can only be imagined but not truly grasped, because the 3rd dimension is in duality, a conditional dimension, you see? But the Ascension brings in the higher frequency, and the true graduation to higher love is thus begat!

This is not to say that you cannot rise above the 3rd dimension and expereince 'Unconditional Love' MerKaBicly in the 5th dimension, in fact many of you do. But it is experienced in 5-D, not in the duality of 3-D, you see? Love is so powerful, more than you understand. Love is the key. Hate only draws more hate. And so we offer you a TRUTH: when you humans finally decide to end war, you will do so by loving peace, not by hating war.

So our message to you is to hold fast! Know who you are, from whence you have come and that what you are doing IS making a difference. Carry the light, and do not fall into despair. Do not feel that you are not making changes. YOU ARE!

Truly from our perspective of higher dimension, your Earth lifetimes are but a flash of time, but an instant of infinity. In the plane of duality the years can seem to bog down, and the burdens may at times feel greater than they truly are. Dear Ones, time is such an illusion. Do not despair!

Thoughts and vision create reality, there is NO other rule ... and visions of light and love are exponentially more powerful than any other frequency of manifested thought within duality. Do you understand?

Know that if you perceive that things are not totally in place as yet for your new role in constructing the new Earth, that they are only a dream vision away, an awakening poised to happen. Dear Ones, if you do not feel this is your home, make it your home. Indeed you are doing just that, and we honor you far more than you realize. We love you so deeply, and we know your doubts, we sense your fears within the filters of duality. We tell you that you have done so much more than you may know. We tell you that you are not alone. The Cosmos is gathered to observe this great graduation, this great dimensional expansion called the Ascension. It will occur!

As Envoys of Light, you are part of a Divine Guild, the Guild of 777: engineers and artists of Ascension, the prophesized 144,000 squared! You can and are connecting to one another far more than you may realize.

Indeed, you are Divine sparks of GOD pretending to be human, and for that we honor and love you so much more than you may sense within the filters of duality. So rise beyond duality, indeed you must! Seek the true divinity within you, the Impeccable Divine Self, and draw from that source, a source based in the dimension of Unconditional Love.

Seek the strength that is available to you; achieve the well being that is your soul right! You are Warriors of Light! You are the chivalrous Knights, the Magdalenes! Strength is not gender based, it is honest and righteous, and it is your nature. It is humble but forthright. You do not impose or attack, but you do defend. True Power lies in the fortitude of Light, in the divine clarity of Impeccability and the higher knowledge of multidimensionality.

There will be a time in the Future, several centuries from now in your terms, when many of you may be asked, "Was there really an Ascension?" And you will smile the smile of inner joy and you will say, "Where there is Light and Love, there is always Ascension." And you will say it from experience, Dear Ones, for you will know that expanding the experience of Light is why you are here in the eternal NOW of this moment.

I am Metatron, and I share these Truths. You are Beloved.

… and so it is. AA Metatron via James Tyberonn

Manifestation ~ Impeccability and the Quantum Crystalline Field

Archangel Metatron Channel via James Tyberonn

Greetings Beloved! I am Metatron Lord of Light, and I greet you one and all, respectfully and knowingly. I greet each of you individually within the exact moment in which you read these living words in the vibratory essence of the eternal NOW. I engage your heart, your beingness, your light, and I embrace you in unconditional love.

Dear Ones, your life, each moment of your physical life is precious, far more so than some of you realize. Far more than most of you utilize. Time is a precious commodity, and it is finite within your duality. Each of you reading these words will at some point in the future transition out of the physical. In your vernacular, you will experience death, you will die. This is a condition of physicality as you know it. Yet so many of you act as though you will live forever. Indeed the soul is eternal, but you will not ever be the same person, the same personality or expression that you are now, in any lifetime or in any other aspect of your beingness.

You are here to learn, Dear Ones, you are here to learn the expressions of your own Godliness within duality, and indeed duality is a gift. Life is a gift. You are here to learn how to cocreate, for indeed you are cocreators of the Universe, of the Cosmos. You are here to achieve Mastery, and so many of you are very close, very near that achievement.

Seize the Day

There is a saying in your times, "Carpe Diem," translated as "Seize the Day." And it is so appropriate. You must seize the moment! So many of you despite your good intentions allow yourselves to be tranquilized into complacency at certain phases or within certain conditions of your chosen sojourns. Many of you waste time, misuse time, and lifetime after lifetime can be squandered.

Some of you say and feel that "Everything works out as it should, all is in perfect order." But Masters, that is a paradox, and like a face card it is upside down either way you look at it. Do you understand?

From the higher perspective all is in perfect order, but from the perspective of humankind within duality, it is not! If it were there would be no need for lesson, no need for what you term as reincarnation. One need but take a look around and know that the plight of humankind on the planet Earth is far from being perfect. Indeed it will NOT work out as it should until you make it so!

The vast majority of humanity however is not yet attuned to its own consciousness, so it is driven by its unconsciousness which is also subject to manifesting its fears and its nightmares. We have told you before that there is a higher light quotient upon the planet now than has existed on the planet since the Golden Age of Atlantis. In terms of pure quanta, there are now enough humans of a sufficient light quotient to bring about the Ascension. Approximately ten percent of humankind carries the minimum light needed for the transition. However, the light quotient during the Golden Phase of Atlantis, while fewer in numbers, was of a higher caliber within the individuals. Indeed there were thousands of Masters of pure light that walked the Earth as energy holders in the Golden Age, and there are far fewer of that capacity and frequency today. So it is incumbent on those of you on the golden path to increase your light and to become greater beacons.

That which you term the Ascension heralded as 2012 is but a beginning, NOT a completion in greater terms. It is a completion of the age of darkness and the beginning of a phase of greater access to higher dimension. Indeed your world will not end in 2012! In fact for most of humankind, December 21, 2012 will seem to be just another day, and many will wonder what all the commotion and hoopla was about. For those of you who view the importance of 2012, we tell you it is the point of wholeness, it is the point in which the higher code termed the Cosmic Axiom becomes more eminent. It is the fulcrum in which Universal Law becomes more whole. From that stance it is a new beginning; it is the emergence of the new paradigm.

It is not a time in which utopia will reign, rather it is the beginning of the potential of integrity within higher dimension. In the remaining years from 2008 until 2012 the Earth is reforming itself, developing a new DNA.

While some of you will sense and utilize the ability to access higher dimension and live within the 5th dimension, most will choose to remain in the 3rd. Indeed it is their prerogative, and within duality "it is what it is," yet it is not perfect!

Creating and Manifesting Your Desired Reality

So we speak today about creating your world! Within the new paradigm there is one reigning truth, and that truth is that all Beings, all souls, all humankind are equally creative, equally sovereign, commensurately free in their potential of mind, body and spirit.

It is an indelible truth that you create your reality. But it is also an indelible fact that ninety percent of humanity is doing so from a vastly unconscious stance. Even those who are able to manifest their dreams often show little wisdom in what they choose to create. In current times most humans who develop the power of will utilize it to manifest ambitions of power and wealth. and thus they create for themselves an even greater lesson and burden to work through.

Your religious texts tell you that it is harder for a rich man to enter the Kingdom of God than for a camel to pass through the eye of a needle. For most that is indeed true, yet abundance is not an evil. Rather it is simply a responsibility, like power that is balanced only with true wisdom; that wisdom is called love.

And love, true love, unconditional love cannot be grasped as yet on your planet by most of you. It can be imagined but not truly acquired, because the 3rd dimension, the dimension ninety percent of humankind exists within is a conditional dimension. True love is a frequency. It is not passion, it is not romance and it is not sexual.

Why You Are Unable to Manifest

There is another passage from your religious text that paraphrased says, "Ask and it shall be given." Yet how many of you truly are able to manifest what you want? Even following the numerous instructional packets available on that topic, how many of you are truly doing it? How many of you have truly, fully manifested the joy and abundance you desire?

We will answer that. Very very few. So if it is so clear, why isn't it easy?

You see, to manifest your desired creation, you must do so within a part of the brain beyond the personality ego aspect. Yet in order to access that part of the brain, in order to access the Divine Mind, one must master the personality ego aspect. There lies the missing key! And Dear Ones, there are no short cuts, there are no bypasses, there are no cheat sheets! There is but one way, and it involves WORK! It requires discipline and devotion, and those must be applied in love! That key is impeccability!

Impeccability

What then is impeccability? Is it walking your talk? What then is walking your talk? Is it doing your best? Of course it is, but what does that mean?

Does it mean taking time to review yourself? Of course it does. Does it mean putting your life under the microscope of critical mind to see what serves you and to see what does not? Does it mean seeing where you have fear and understanding why?

Dear Ones, this is not meant to make you feel guilty. It is meant to make you see why you are here and why you are not able to manifest your true desires! It requires achieving the crystalline vibration in your daily lives. And achieving the crystalline frequency in your daily life is called impeccability.

But you cannot do that if you do not even take time to see yourself! You cannot achieve this if you rest on your laurels! You cannot do this if you place yourself in victimhood! You cannot do this if you do not love yourself! You cannot do this if you hide all your bad habits under a rug of ego image projection.

Dear Ones, you cannot fool yourself, and you cannot fool Creator God, of which you are an intricate part!

Now, we are not telling you this to make you feel guilty or to make you feel inadequate. Indeed you are already on the cusp of this Mastery, and merging your vibration into the Quantum Crystalline Field is indeed within your grasp.

Within the Crystalline Field is the Truth of Universal Law, and that Truth is within each of you. Within that field you are already PERFECT, and Dear Ones, within multidimensionality you are already there. But to consciously merge your integral beingness into that field, certain vibratory requirements must be met.

Within all of your Beingness is the Truth, and it exists as a point. Within you, within the crystalline aspect, that Truth is now expanding. It is now coming forth as a challenge. Your Quantum SELF is within perfection, and every aspect within your consciousness that is currently less than that, less than whole, is being confronted. You see, anything less than whole cannot pass through the frequencial filter, the vibratory crucible of the Quantum Crystalline Field.

Everything that you carry in your consciousness that is not in integrity is being engaged. And you cannot hide it, you cannot smuggle it through, you must face it and dissolve it.

Is it difficult? Yes, but you can do this, you must, and sooner or later you will.

Self Review

Achieving the Quantum Crystalline state of clear mind is not easy, but it is not as hard as you may think. It is letting go of that which does not serve you. It is having a clear conscience, a mind that is guilt free, self loving and fearless. It is walking your talk and getting rid of anything that prevents that.

So which aspects of your life are wisdom, and which aspects are bad habits? Have you taken time to review these and to list them? And what about your fears, your angers, your complacency, your victimization, your guilts and your doubts? Which elements compose these blockages, and indeed they are obstacles that keep you from achieving the final step into Self-Mastery.

Most of you reading these words consider yourselves to be 'on the path.' Many of you attend seminars and workshops studying metaphysics, listening to channels and striving to grow, to teach, to be a beacon. And when you gather together something wonderful happens, something called higher consciousness is formed by the very mass intent of the group energy that gathers. You experience an extraordinary energy, and you feel exquisite, you feel the energy of 'home,' of 'higher dimension,' and you bath in it, you experience bliss within it! Is this not true?

And so we ask you, after you leave the seminar, after you leave the gathering, how long does the blissful energy stay with you? How soon after you go back to your jobs and everyday life does it take to wear off, and you experience again the mundane, the stress, the depression and fall into bad habits?

What happens to the energy you leave with? Where does it go?

You see, within the group energy, within the seminar you put forth your best self, you adhere to the lessons of higher thought, you treat everyone including yourself with love. You give love, you allow yourself to receive love, you meditate in groups and you create a mass of exquisite energy. You adhere to the walk of the higher self! You achieve the crystalline frequency of clear mind.

But afterward one week passes, then two, and the magic begins to disappate. Two weeks after the spiritual gathering, how much of your day is devoted to love, to joy, to meditation, to applying the disciplines that you learned, to doing your best? How much of your day do you disallow victimhood and depression? Is it ten percent, twenty, five?

The Importance of Detachment

Now, within your brain is an aspect beyond the personality ego that we term the detached observer. This is not the inner narrator, indeed the inner narrator is 'attached.' The inner narrator is attached to the emotional body. This is not the detached observer. Quiet the inner narrator as if you are in meditation, and allow the quiet observer to preside. Do you see, do you understand the difference between the two? The detached observer is in a different part of the brain and is beyond the personality. It is beyond judgement; it simply observes from a stance of detachment. This is important. Develop this energy, develop this presence, this state of being.

Now, take one hour twice a week before going to bed and allow the detached observer to view the actions of the personality for the past two to three days. How do you observe objectively what you experienced in your days in the seminar, and how does it compare to the days you have spent since?

How many of you can see what it is you need to improve upon? Is it easy to see? It is, if you work at it. Is it easy to work on? Yes, no? Indeed Masters, it is NOT easy! It is not easy because most of you choose to put aside that which is difficult to face, that which you deny, that which you cannot admit. And you put it on a back shelf or sweep it under a rug, because it is in conflict with who you want to be. You hide it from others, and you hide it from yourself. It is very hard to engage those difficult aspects of your own shortcomings. Yet it is these very habits that must be faced if you are to achieve impeccability.

Hidden Ego: Spiritual One Upmanship

Many of you are leaders and teachers, and you have worked long and hard to acquire an advanced level of knowledge and spiritual truths. This is admirable. However, with advancement comes the need for humility. How many of you have attended metaphysical gatherings only to be somewhat put off by someone who makes it a point to let you know how far they travel in their dreams, how many great visions and celestial teachers they see in the higher realms; the implication being that they are advanced in mastery and want to be sure everyone knows it. This is an understandable pitfall, but is not the walk of integrity. Spiritual 'One-upmanship' is rampant in New Age circles, yet most offenders would not even be aware they are coming across in such a pompous manner. There are many covert trappings of ego within spiritual circles. No one is exempt. Leaders and teachers who advance to a certain level are particularly susceptible to this trapping of ego when they begin to achieve a following. Most that have it do not recognize it. Are you one of these? It happens to most at some point, Dear Ones. Humility is the way of the true spiritual leader, yet it is so easy to blindly fall into self aggrandizement.

We tell you in love, Dear Ones, that it happens to all of you. It happens in all of your lifetimes, the move and shift day to day, in and out of integrity. Within duality, whether you are a teacher or a student (and all of you are both!) none of you are above it. And that is why it is necessary to recalibrate through the infinite mirror of truth, of self review within the detached observer within self. There is such a fine line between self love and ego. The paradox is that the former is a necessity of impeccability and the latter is a hindrance to it.

On the path of Mastery, you will be forced many times to choose between love and power, and power is very seductive and often well disguised. Pride comes before the fall. Many have risen to great heights in spiritual achievement only to lose it all by becoming blinded by ego.

Self Review and Calibration

And so let us review some of the finer points around the ongoing need for self review and how this can assist you in achieving impeccability. How many times during the week or month do you have the experience of telling a small white lie, and how many times do you do it? How many of you feel that it is not as bad to tell a little fib as to tell a big whopping mistruth? From a vibratory resonance, is there a difference?

How many times during the month do you show up for your job a little late or leave a little early? How many times during the month do you have an opportunity for forgiveness, for mercy, for kindness to a stranger but look the other way? How often do you criticize others and involve yourself in what is termed gossip?

And for those of you who spent lifetimes in the cloth of Christianity lying prostate asking for forgiveness for the dogma of original sin, have you condemned yourself finding it easier to give than receive, finding it easier to give love than receive it because you feel yourself unworthy? That is outside of integrity too. That is not impeccability. Self love is not a gift, it is a requirement! It is not something you earn, it is a frequencial requisite.

Sexual Integrity

Sexuality is another area that is very complex, very confusing for many of you. Throughout the eons of time there have been many systems of belief, many varied expressions, experiments and modalities within the various cultures across the planet on sexual expression within biology. Some were very inhibited, others extremely liberal.

Most humans in your present society have conflicting issues about their sexuality and sexual expression. It is not ironic that many of you will achieve the highest balance in your lives only after your body ceases to produce sexual hormones.

Yet the release of kundalini, of chi through the sexual act is among the most powerful energies available to mankind. It has been used for good and has fallen into wasteful misappropriation. The key is intent.

When there is consensual attraction teamed with caring and mutual respect it can be exquisitely sensual, beautiful and sacred. It is capable of transporting the spirit to higher realms, combining the trinity of body, mind and soul. When this occurs, particularly among members of the same soul guild, there is an energy imprint imparted and a synergistic effect that is beneficial on many levels.

Sexuality is the natural expression of love that flowers between souls. It can be the highest natural expression of love between people, regardless of gender. Yet it can become a source of guilt, a source of control and a source of bias and judgement. It can become a source of self aggrandizement and addiction, particularly among those of the male gender. It is often misunderstood and often misused. Relationships should be based on agreement, and the freedom of this expression should be honored, yet optimally aligned with the higher chakras.

High intent within sexuality is extremely beneficial on all levels: physical, mental and spiritual. It can be a source of rejuvenation and regeneration. Yet some choose sexual expressions that border on debauchery, manipulation, greed, conquest and misaligned addiction to physical orgasm.

Without judgement, we tell you that sexual energy is a gift of exquisite energy, and when used in higher intent it offers a glimpse into the sacred orgasmic bliss of the celestial realms. It is life force and should be used wisely, encompassing the body, mind and spirit. Its responsible enfoldment with high intent, sharing and love, Dear Ones, is another aspect of impeccability.

Strength and Focus

Oh, you say, this is so complex! Of course it is! But it is why you are here, Masters, and you can do this with self review and strength of will! We are not asking something of you that is beyond your ability. We are not asking you to do something that you have no innate understanding of. Indeed you know these truths already! We are revealing what your ego has attempted to hide.

So what is this we are revealing? It is no mystery, Dear ones, it is the crystalline frequency. It is the resonance of impeccability. It is the mental crystalline resonance, and it takes effort, it takes will, it takes work, it takes time and it requires humility! Masters, you are on the cusp of something extraordinary! Seize the day, do not waste precious time, do not waste another lifetime! You are so close to acquiring a dream that you have sought for many many lifetimes.

Everything you have hidden away, everything you have ignored, every little obstacle you have put on the back burner will come forward and give you opportunity after opportunity to resolve it. It will not go away.

Unresolved Energy Blockage

All unresolved energy must be transmuted before you can step into impeccability. And until it is faced it will surface again, time after time after time until you confront and dissolve it!

Unresolved energy is that which you have left in chaos in your life, and some of you on the cusp of mastery have reorganized these little bits of chaos and repackaged them into tidy little boxes that you have justified in one fashion or another. You have justified them, disguised them, hidden them and spun them into unrecognizable forms that allow you self exoneration. That is called denial. It is called postponement, and you put it away to deal with in another lifetime. That is called wasting time, wasting opportunity. The unresolved energy you repel has been polarized, and it sits outside your mental field disowned, disavowed, denied; ticking away. So you work at maintaining the status quo, charging forward. Some of you on the threshold of Mastery think you can move forward irregardless, and this, Dear ones, is not the case.

You may feel saintly because of the many gains you have made and the many issues you have resolved. But in most cases, the things you have resolved were resolved because they were comparatively easier to handle. The ones that were really hard, really hidden, you have not dealt with.

Polarity Physics

You see, Masters, the closer you get to light, the stronger you attract the dark. It is true in real sense physics, electromagnetics. Pure positive energy has the greatest attraction to negative energy. So you must have the wisdom, the humility, the strength and discipline to deflect it. There is another law of physics that you must understand, and that is when you achieve the crystalline vibration called impeccability you rise above magnetic duality attraction and enter a field called zero point. That is blissful detachment.

Now, hard energy, hatred and anger is another piece of the puzzle. How do you deal with this? This is not a great mystery to you, in fact you created it. It is a part of duality. It has two aspects. Your bible talks of turning the other cheek. What does that mean? This does not mean you apologize when someone steps on your foot. Part of the paradox is indeed standing up for your truth. But it does mean you do not step on the feet of others, intentionally or otherwise. Do you understand?

Aggression is feared by many of you yet in its core, aggression is one of the most creative forces on the planet. The force that enables a small seed to burst forth through the soil, raise its tender green shoot toward the sun and burst into a beautiful flower is pure aggression. When human conflict occurs, aggression in its pure form is often the means by which humans learn more about themselves. We are not referring to a war, but it is rather when both parties are standing in their truths and finding the means to fully express their views. Conflicts are often 'contractual' setups, opportunities that teach both parties.

Each of you have an opportunity to stand in impeccability within any conflict. You can deal with conflict without engaging it. Do you understand? Deal with it, face it from a stance of emotional detachment as the observer, and that is not easy yet it is the way of the Master.

Each of you has an opportunity to be impeccable every day. The scenario in which you recognize your own failings, your own conflict with integrity is the day you encompass impeccability, and indeed it is a journey. Likewise, the day you stand in your truth with willingness to recognize another person's truth, you encompass integrity. There is a saying that the right for a man to swing his fist ends where another man's nose begins. Impeccability involves recognizing the rights of others without surrendering your own. Be good to one another, measure twice and cut once. When you make a mistake, own it, forgive and be forgiven, and do so within detachment.

Clarity as a Vibrational State

There is often the feeling that comes when facing the conflicts and difficulties of life that it is too hard. Many feel at times that it is just too difficult. But Masters, it is never too hard, it is never too difficult. The moment you accept your challenges and rise to them in detachment is the moment you achieve clarity of purpose and integrity of clear mind. This knowledge, this moment of clarity then becomes the power driver of your life. Irregularity is removed. Within this detached state of crystalline lucidity you are able and only then able to access the divine mind, the multidimensional brain: the extraordinary state of clarity that has always been there, but never quite accessed in emotional states of thinking.

It is the divine mind, the super subconscious, that contains within it the aspect of man that is divine, the God mind. It is achieved through impeccability of clear thinking, of lucid mind within the extraordinary crystalline field. At this point, the human is in a crystalline geometric field of the Merkivah. The human has become a divine thinking soul and has crossed the veil into the creative divine mind.

The divine mind is only achieved, only accessed through crystalline resonance within crystalline thought waves. Crystalline thought is above emotion, above petty feelings. It is achieved in detachment. It is the crystalline lake of Shamballa, of Nirvana, as smooth as glass, no waves distorting its mirrored visage.

Masters, when you achieve this clarity, this level of unbiased focus you have a direct alignment to the extraordinary divine field within you. This field is crystal, it is unclouded, unwaved with doubt, unfettered by fear, victimization or any of the petty issues that clutter the day to day thought patterns of the limited ego mind. This field of crystalline thought, divine mind is devoid of the turbulent tempestuous raging of victimization, judgement, lust for power, control, anger and passionate involvement that is so often imprinted in the repeating dialogue of the inner narrator of the ego mind. It is detached from and above the storm of emotion. It is bliss; it is achieved in impeccability.

This state of beingness is the level in which we achieve the creative divinity of life. This state of BEING is capable of manifesting all that we desire within wisdom and LOVE, the frequency of LOVE.

Clarity as the Requisite for the Crystalline Creative State

Everything that you have read and been taught about how to achieve impeccability and how to manifest joy, health and abundance in your life can only be achieved in the detached state of the crystalline mind. We say again and again, it does not occur in the attached state of the ego mind, and it never will. All you have read and learned about creating your world and manifesting your reality is merely an unachieved potential and an intriguing philosophy until you experience it directly, until you do it.

You cannot and you will not become a true Masterful Divine Being until you decide to become impeccable, until you decide to make the effort to review your life and remove the hindrances. Until you love the self, then and only then can you truly love others in unconditional higher dimensional love within the crystalline field.

Seize the day! If you are hurting others, if you are caught in ego, in victimization, control of others, depression, a lust for power, STOP IT! Change now! Until you do, you are doomed to repeat the cycle.

This is very achievable. Simply put, it is being tapped into the paradigm of Universal Truth and of doing your best. Drop those habits that do not serve you and live in the clear consciousness of that goal. That is impeccability.

If you make a mistake, that does not condemn you, there is no condemnation. When you err, simply recognize it, learn from it, make the adjustment and move back into the crystalline resonance of integrity.

There is a Divine Intelligence, a lucid clarity that takes you to the Mastery you seek. That is available, and many of you are so close to achieving it. Don't stop, clear the closet. The lucid mind, the crystalline field is achieved when the inner narrator is stopped and the frontal mind lets go of issues.

- *Seek the Light*

- *Be Honest*
- *Review Your Life and Calibrate*
- *Put Your Best Effort Into Everything You Do*
- *Encompass and Live Your Beliefs*
- *Remove Fear and Doubt*
- *Remove Anger and Control*
- *Love Yourself*
- *Love All Life*
- *Nurture Others*
- *Be Detached*
- *Create Your World*

There is no ecstasy more supreme than achieving the Divine mind. It is why you are here. Sooner or later you will learn this, and the sooner you do the sooner you will find within yourself that elusive doorway into the realm of Creatorship, of Mastery. It has always been there, and it awaits you, Dear Ones. Your higher Self is already there.

I am Metatron, and you are beloved.

… and so it is. AA Metatron via James Tyberonn

The Infinite Geometry of Sacred Sites ~ Quantum Crystalline Activation

Archangel Metatron Channel via James Tyberonn

Greetings, I am Metatron, Lord of Light, and I welcome you.

Beloved Masters and Earth-Keepers, know you that I Lord Metatron greet you at a geometric concavity and precise angle above your space time, in a dimensional corridor of your choosing, one that varies with each of you receiving this message. Dear Ones, you are not here by accident! This message is given to each of you in a precise point of intersection. It is time to activate your crystalline field.

So indeed, we welcome and acknowledge each of you within your individual vector in a space created for each of you in golden spiral as a unique pattern, because it is after all geometric patterns that are significantly of interest to you and to this discussion: how this energy flows and arcs, how it aligns itself geometrically with this point, this geology and that angle and how one grid and ley line juxtaposes with another. Accordingly, we will tell you that many of you Earth-Keepers have chosen merkabic EMF lattices that are uniquely capable of flowing in a spiraled pattern with the crystalline activation necessary to this time of ascension. If it is that you are a spiral, we would say that you are mid stream, both within and without the ascending spiral. And without the spiral are conscious pristine angles. You see, if you approach such understanding as a destination, dimensionally it implies that you approach it from more than one side, and for that we applaud you. Therefore the angles become more distinct and unique to one another – more acute as you near your perspective, your interest and crystalline destination. More acute then becomes the interest, more specific then become the questions, more fine tuned the discernment that fulfills and defines you.

Indeed Beloved, I tell you that whatsoever you seek you will find! But be prepared to work devotedly for this knowledge. When you choose crystalline activation, the realization will depend on your level of awareness, your light quotient and your will to become Divine. Then you will serendipitously discover that knowledge incorporates itself in harmonic coded angles into your beingness and reforms with what you already know. Knowingness becomes wisdom, and wisdom refines itself in clarity. Progressively from the clarity of understanding comes the projection of divine interest, of importance both to self, to others and to the Cosmos. You see, you are not separate from your Cosmos, you are indeed part of your environment, an environment horizon that is crystalline!

And so Beloved Earth-Keepers, please consider the questions posed by the channel and the answers that are given herewith. Consider the information that has been stored patiently until this moment, for there is much to know, much to gain access to, much to bring clarity from and much yet for you to do. Indeed, perhaps other sessions will be described at later intervals for those who desire to uncover not only the truth but the deep understanding of same.

Now, the channel asks about these areas of sacred sites on your Earth that he refers to as powernodes and sacred sites. We tell you that humankind since the beginning of Earth sojourn has recognized certain areas of your planet as points of power, areas that resonate with a higher frequency. Those areas on the Earth which you term powernodes and sacred sites are apexes of infinity. These apex points are the primary dissemination mechanism of the higher dimensional energies of the heralded

Earth Ascension. It is indeed through what may be termed vortex-portal templates that the frequency of Earth is maintained, adjusted and amplified. That is why humankind has always been drawn to them, for humans are also amplified, clarified and discover doorways to crystalline beingness in their energies.

Now, it is important as a prelude to this presentation that we examine the greater reality of the fabric of your universe. The Earth, the universe and indeed the cosmos is composed of living conscious energy ... and that conscious energy consists of geometries, light and electromagnetics.

The Earth is multidimensional, and so are you. Twelve dimensions exist upon your planet; 356 dimensions reside in your Universe. Geometry, sacred geometry, is the mosaic underpinning of them all. I, Lord Metatron am the generator of the fabric prima-maitre of this underpinning, and you are the weavers in a vast agreement that incorporates more of your multidimensional consciousness than most of you understand. Accept this truth for now as a postulated axiom.

The channel has written about the great Greek mystic and philosopher Plato, who described your planet in a moment of lucidity as looking like a ball with twelve sides when seen from afar. What he was referring to was the dodecahedron. The dodecahedron is the inner triune geodesic geometry of the three in one grid complex that encompasses your planet. We have described this to you previously in detail. This grid system enables and facilitates your experience on the planet and the fabric of reality in which you live. To put it in your computer vernacular, the 144 grid system is the Windows program that enables all humanity to have this unique Earth adventure expanded within duality. The 144 grid system is the server to your mental programs in and above duality.

Before the advent of duality, your planet was governed and protected by the Divine Firmament. In this phase, the Earth was devoid of duality in a non polar field we refer to as point zero energy. Before duality humankind was androgynous on the Earth. The Earth was a living library with all Universal knowledge present as a protective vibratory field, fully accessible to all. The Earth was like one of your ornamental glass covered snow globes; no matter how it was shaken nothing was dislodged, it all settled quite softly. But it was knowledge without experience. Mankind chose to leave the shell and experience knowledge through growth in choice, in duality. Your religious texts, skewed as they are, referred to this period in the metaphor of the Garden of Eden. The dissolution of the Firmament begat the planetary field of duality and birthed the formation of the grid system. Duality in the metaphor of Eden is represented by the Tree of Knowledge. The forbidden fruit of knowledge, the infamous eaten apple represented man's waking sojourns into duality and lesson. Adam Kadmon split from androgyny into dual polar halves, male and female, and the serpent represented the chakric kundalini in its application for human sojourns into polarity, duality and lesson. Mankind chose the Earth experience, put on filters, and thus it began.

Adamantine Essence in Sacred Sites

Now, we have told you that the firmament was composed of two elliptical energy fields around the Earth, one electrical in nature and the other magnetic. One emanated the energy of crystalline platinum, the other of crystalline gold. The two created a synergetic crystalline field that permeated the Earth and allowed for ontological, fully aware consciousness in a unified, multidimensional state.

The dissolution of the firmament created a crystalline energy particulate, an etheric plasma that was anchored and stored within the mineral composition of the planet and indeed within human DNA. This subatomic particulate within the crystalline fields of the planet is similar in nature to what is termed Adamantine Essence or Akash. This vital essence is an etheric life force and is in more plentiful abundance within areas you call powernodes, or sacred sites. Elixirs of this potent essence indeed were mined by the Atlanteans and used in the rejuvenation process within the Temple of Healing on the Atlantean Isle of Poseida in an alchemy process. It is now being generated in greater mass and projected into powernodes and sacred sites from within the living Earth as a result of the tectonic movements.

Adamantine essence, Akash, is a multidimensional etheric particulate that has been called by the alchemical ancients as the elixir of life. Adamantine occurs in fields of natural mineral crystallization and is both telluric and etheric in nature. These fields do not exist within the field of the grid system as an evenly distributed component. But it does occur in all of what you would term planetary powernodes. All sacred sites contain to varying degrees etheric fields of adamantine resonance, and as such they are elements of the crystalline fields that are amplified and enabled by the grid system. Do you understand?

Sacred Sites

Sacred sites in all of their forms are of great importance on the planet, and this has always been so. As we have said, they perform a myriad of tasks including balancing the planet and connecting the Earth to higher dimensions. In addition to being the receptors and distributors of the energy of the planetary ascension, they are also key accoutrements of mankind's ascension.

A sacred site can be defined as an infinity point which is of relevant importance to both the Earth and to humanity, to within and to without the Earth, to the celestial heavens and to the celestial body itself. It is that which empowers, it is that which evolves, it is that which affects conscious evolution, it is that which is destined to remain a part of humanity's cellular existence. Indeed the adamantine essence of its energy is also within your DNA and synergistically melds with humankind in fully activated sacred sites.

Now, in specific points on your planet the 3 in 1 crystalline grid, the dodecahedron, icosahedron and double pentadodecahedron, are unified into concentric lenses that allow for all 12 Earth dimensions to occur in a more tangible manner. This creates a phenomenon that the channel refers to as zipped space or harmonic dimensional overlay.

When you as humankind enter into these special areas, you become immediately affected. Your energy increases, your awareness is enhanced and your multidimensional aspects are more tangible. The degree of personal expansion depends of course upon the awareness level or light quotient of the individual, but all visitors are affected to one degree or another. Those seekers who enter into these areas in reverent spirit are given the opportunity to step up their frequency, and they have an extraordinary opportunity for multi dimensional experience.

The channel has classified in his study a grouping of five categories of sacred sites:

- *Natural Telluric Energies: Mineral Deposits such as gold, quartz and diamonds, mountains, rivers, oceans, volcanoes, rivers and springs*
- *Man Made: Pyramids and Cathedrals Built with Sacred Geometric Architecture*
- *Grid Points: Points of Grid and Cosmic Alignment*
- *Sacred Plant Life: Fields created by oak, redwood and*

other powerful plant life systems

- *Human Imprint Sites: Such as Tombs of Great Leaders and Sites of Historical or Religious Events*

We will confirm this in broad terms. However, it is necessary to add the caveat that a powernode is, in Metatronic terms, the raw basis for a sacred site, but not necessarily a sacred site. Humankind can energize and groom powernodes into becoming sacred. Let us say that a powernode then is an energy field that resonates to a higher frequency, originating from a geologic or telluric source. It may be a mountain, valley or river confluence that exudes an amplified sense of balance, well being and tranquility. When it is inhabited or frequented by people who honor that land and meld their acknowledgement and reverence energetically through prayer or ceremony, then the powerpoint can become a sacred site. In that way their energies and high intentions are melded into the energy of the powerpoint, and it evolves into a sacred site.

Now, in terms of Human Imprint Sites as sacred, we will say that the energy of a powernode can be amplified but not necessarily created. There must first be the raw energy there. Now indeed as the channel has classified, when what you might refer to as a great historic event occurs in a location and multitudes of people are thereafter drawn there to visit the location of the great event, the energy they amass can groom the site to a higher frequency, and a sacred site of sorts will evolve by amplification. But again we stress that a potent energy potential was there first.

Powernodes

Most powernodes emanate their initial energies from a blend of geological sources. Metamorphic rock, igneous rock and to a lesser degree sedimentary strata and specific water sources are the foundation of most of your telluric geological powernodes. Particularly metamorphic mineral formations from what is termed the Precambrian period are laden with Earth pegmatite treasures. Crystallization within metamorphic and igneous rock occurs from the eons of pressure that transform any rock form into its pristine crystal metamorphism. You see, even those termed the rock nation by the indigenous have their progression of growth to crystallization through pressure and time!

Now, our current science theorizes and purports that your Earth and solar system are approximately 4.54 billion years in age. However, your Earth's oldest rocks cannot truly be tested because they do not exist, they have been ground by the pressure of the ages, destroyed by the cyclic process of plate tectonics. We will say that the calculation of Earth age is only functional in 3 dimensional terms. Your current science is only capable of study in linear time and is not currently capable of understanding that perspectives of science change in multidimensionality. Only now are you reacquainting yourselves with quantum physics and the validity of parallel dimensions. Circular time exists to you only as an enigma of grand intrigue. We tell you however that within certain powernodes and certain sacred sites, multidimensional and parallel inner and outer worlds indeed occur and, in specific space time sequence conditions, become quite accessible.

Nonetheless, your geological understanding of rock phases and crystallization is quite accurate as applied to our study of powernodes. Areas of pure mineral fields, metamorphic rock and igneous rock are the most potent sources of mineralogical energy fields. The noblest of your Earth mineralogical energies are from gold, platinum and all crystalline minerals and compounds. Powernodes not only emit an electromagnetic field, they also are capable of bending, refracting and dispersing light in myriad ways. They are unique in their abilities to affect the time space continuum, especially when twinned with specific celestial arcs, angles, electromagnetics and gravities.

Grid Geometry of Sacred Sites

Now, the channel asks about the connection of sacred sites to the grid. Sacred sites are not independent of the grid, however they are not fully relative to it currently, you see? This is a work in progress, we will say, a work many of you Earth-Keepers are actively involved in doing. Now, the channel has noted in his first book that some of your scientists have attempted to find a specific geometric pattern and grid alignment with your more notable sacred sites, and this sadly has not been successful. The points they connected did not form the dodecahedronal pattern they sought. However, they were on the right track. The seeking of geometric patterns in sacred site alignment is the appropriate concept, but your scientists do not have all the pieces of knowledge required to solve this equation. When one attempts to align to the grids or sacred sites now, they do it upon current knowledge of which very little is understood of the grids. Even less is known of the technology of ancient sites from the time they were erected and brought into alignment and full power. In those times a different geometry was in effect. In most cases, the Earth was at a different axis angle. Literally there were different stars above at that time, and these were and are rotating, changing their position, no longer in the original power angles, you see? Indeed in ancient days the sacred sites of the planet were in a fluid geometric alignment with the grid. Additionally, over the millennia of passing time many of the ancient sites are now inaccessible and thus unknown to you. Some are now below the sea, others destroyed by plate tectonic shifts and many more under the sand and silt of ages. We tell you that below the sands of Egypt, Ethiopia, China and Turkey remain magnificent pyramid complexes whose existence is yet to be imagined by your anthropologists. Two of these are still quite active. So you see why the geometries cannot function in your now?

And so the patterns that are being applied for these theories do not fit as they once did in ancient times. In order to truly understand this, each sacred site would need to be attuned to the height of its potency at the time it was acknowledged, energized, erected, honored and aligned with each grid at that time. This process would, in eons past, superimpose each site geometrically with one another and then synergistically correlate this into the planetary and stellar grid systems. If this process were understood, a distinctly functional pattern would emerge. In the present now, there is no one being, no organization upon the Earth that is fully capable or prepared to do just that. There is no one computer that will compute it because it requires a consciousness, a knowledge and awareness that is not yet available.

Now, sacred sites relate to one another in distinct modalities. Some sacred sites, such as Lake Titicaca in Og (Bolivia and Peru) and Mount Shasta in California to site two such examples, are mega site-satellite complexes that contain a multitude of balancing powernodes within a distinct area. These energetically circulate around a center core and as such fully balance the energy into an enormous sovereign energy system. In smaller systems, sacred sites will triangulate. All sacred sites will communicate through energy systems based on harmonic oscillation, ley lines, meridian lines and etheric axial tonal lines.

The crystallized matrix of sacred sites projects a specific harmonic wavelength through vibratory energy pulses that communicates through harmonic energy oscillations. That is a scientific fact understood by your academics. It is called the Law of Harmonic Oscillation. For example if one has a tuning fork in the key of C, and then one plays a C note on a piano, the musical vibration from the piano will provoke vibration in the tuning fork. The same also occurs in frequency oscillations between power points on Earth and in higher dimensions. Therefore, a sacred site or powernode with an energy source composed of strata from Precambrian crystallized granitic gneiss, for example, will emit a frequency that is capable of effecting and communicating with all formations of identical type regardless of where they are located on the surface of the planet. The same applies to sacred sites in terms of their grid and celestial alignment frequencies and their geometric projection. Sacred sites will also create axial tonal attraction linkage based on their divine purpose. Certain mountains and volcanoes project what can be loosely termed as male and female energies. These again will connect on a basis of polarity attraction. Do you understand?

Those who truly grasp this can understand the complex nuances and differing aspect of power held in different sacred sites. Again, these systems and points have shifted over eons of time, but those who are adept can locate this shift and realign these systems to the grids. But those scientists who have attempted this from an academic stance in recent times are not truly prepared to do so. It is quite amateurish how they go about it in truth.

Some sacred sites are aligned to planets in your solar system; some are aligned to stars and celestial bodies in other systems and galaxies. Machu Picchu for example, which is a satellite to the Lake Titicaca complex, is currently aligned to the planet of Saturn. But the alignment to stellar grids and distant systems is done through and by means of the lens of your 144 planetary grid.

Not all sacred sites are equally embellished. Not all sacred sites are yet fully active in the sense of a conscious mega complex, because not all are intended to be. When a sacred site is fully conscious and fully active, it is connected through all of the components and layers of the 3 in 1 grid system. A fully activated system of sacred sites contains mega centers, satellite nodes and for lack of a better term sub powernodes. Accordingly, if one sacred site were to become damaged or obsolete, others would be robustly activated to take its place. If there would be a shift on the axis of the Earth, for instance, it would cause the planetary and sub planetary chakras to realign, redesign themselves, reactivate themselves and thus relate to the grid differently.

Now in the coming decades, understanding of the geometric alignments of sacred sites will be more clearly understood. This will occur primarily with multidimensional assistance. This knowledge will detail how sacred sites are aligned to differing celestial energies, systems beyond the 144 grid. The geometries of hyper dimensions distant from your solar system incorporate the geometry of light. It is formless geometry which cannot be understood in your current level of comprehension. Beyond your Earth, coordinators of star systems utilize a navigation of light geometry. In distant star systems the grand crafts are navigated based upon this geometry. You see, if one truly understood the stars, whether they pull or repel and what is their gravity field, then one could align a specific design, a geometric operating system that would create a source of power that could propel life through sheer geometry. And so the star ship can travel through space via a very unique design. So we tell you that light geometry is not only another form of navigation but also a source of immense power. Light geometry varies with velocity and dimension, you see. Your visible light spectrum is but a candle to the geometric light of the Cosmos.

Teachers via the Third Language

Sacred sites are conscious teachers, and they teach in the third language, the unspoken resonance of light. The Native

Americans understood this and always made an offering, asking permission before entering a holy place. When a pilgrim enters a sacred site in reverence, an energy envelopes the auric field, and one's entire being is shifted. One learns through stillness, through quieting the inner narrator and allowing the wisdom to flow inward. The wisdom of 'Universal Knowledge' is activated within your 'God Mind' and within your very cellular DNA. Certain types of sacred sites, primarily those that emit light through an octahedronal geometric overlay, purify you. They literally force you to face your imperfections and parts of your life patterns that are in conflict with the Divine Self. The seeker becomes purged through the crucible of deep self examination. Other sacred sites, those that project light through the dodecahedronal lens such as Roslyn Chapel, facilitate your journey into higher realms. These are termed stargates. Indeed, Scotland's Roslyn Chapel houses one of the most potent stargates on your planet ... and you thought its mystery was about the elusive Holy Grail!

Now, some of your metaphysicians speak of activating, healing and anchoring sacred sites. While we understand the intent of such descriptions, it is in most cases the other way around. It is you who are activated, healed and anchored by the heightened energies. Seekers do imprint their energies into sacred sites, and that energy is amplified and does enhance the vibration of the matrix. But the energy you feed into sacred sites is returned to you exponentially within the quickened influx of higher energies and adamantine particulate matter occurring in powernodes.

In times past guardianship of certain portal vortex complexes were necessary, particularly during your middle ages in Europe and parts of the western Mediterranean when secret societies undertook such roles. Secret societies of Gnostic savants infiltrated the Catholic Church through priesthoods and guilds. In doing so they played influential roles in not only preventing the destruction of what were termed pagan sacred sites, but they actually had monumental temples and cathedrals utilizing sacred geometries built precisely over them to amplify them. You are more indebted to these brave souls than you may realize, Dear Ones.

However that crisis has passed, and this role of guardianship has now transferred to one of stewardship. Even this is somewhat misunderstood by many of you. While at times certain residue energies or imbalances may need clearing, what is generally required of mankind is simply to recognize and honor these sacred sites and powernodes in intent and ceremony for the magnificent beacons of light that they are. The valid aspect of stewardship in the present is to insure that sacred sites are acknowledged, preserved and not commercially destroyed. How sad that many of the great red rock temples of Sedona are now overrun with shopping centres and the like. Masters, the Earth is willing to sustain you and willingly concedes some of her treasures, gems and resources to you, but they must be utilized with care. While many of you may not think it so, the Earth is in cognizant balance, and nothing is being taken from her mineralogical resources that she does not willingly give. She retains her balance. It is so. In the decades to come, your industrial exploitation of hydrocarbons will diminish, a victim of its own greed. Cleaner, more economic renewable energy sources will come to the forefront. The greatest travesty in current times is the ruthless harvesting of old growth forest and your sacred ancient redwood forest. Once these are gone, that same energy cannot remanifest itself for millennia. That, Dear Ones, is a great travesty.

Vortexes and Portal Systems

Now, the channel asks about the vortex-portal system. We would define a portal as that which allows one source of energy or beingness to transfer its energy to another source, from one world to another, such as physical or non physical. It allows that which is a transmission, transition and transmutation of energies for purpose: some portals through dimensions, some through space, some through time, some through stellar arcs, some through all of these. Some portals are direct gateways to specific star groups, some portals are wider stargates to the cosmos. The adept can determine which ones are of which type. The channel is well on his way to rediscovering that ability within himself. Such knowledge you have had in previous experiences, and it is being activated by your travels to and from, here and there across the continents of the globe. Many of you in the past three decades, in addition to the channel, have been compelled to visit sacred sites. There is great purpose in this. Do you know why?

It is because each sacred site gives you an imprint of its unique message, its unique geometry. Can you imagine that you carry within your field the energy of every sacred site, every power point and every grid point on each continent you have visited? You all have the ability to connect them to yourself and to one another, Dear Ones. Can you imagine that you might help connect them to the evolving grid and spread the message of how this can be done? Is that not what you Earth-Keepers have chosen to do?

In every sacred site the door to sacred knowledge is opened, and a catalystic energy exchange occurs. You become imprinted with the keys of frequencial wisdom unique to every sacred site you visit, and in kind you imbue your imprint into the energy of place. You receive light energy from the site, and you feed energy back into it. A synergetic transposition ensues as each seeker begins to manifest the impeccability of every sacred site visited, each melding and building upon the other.

You see, every time humans enter into a sacred site, it becomes part of them and they are changed. This builds with every unique site visited. And when you actively involve your ceremony of connecting the sites, one to another and to the 144 grid, indeed you and the others who do so accomplish a great deal more than you imagine. You are connecting the dots, in your vernacular. You and many others are laying out the geometric pattern by incorporating your energies into that of many powernodes that will indeed become sacred sites. Many of your UNESCO Sites, Heritage Sites, National Monuments, National Parks and State Parks are in fact sacred sites, but they are in modern times only considered recreational or nature preserve areas and not yet utilized or acknowledged as sacred sites. Salar de Uyuni in Bolivia is one of these, you see. It is one of the most powerful nodes on the planet but not acknowledged or understood for its great healing abilities and dimensional access. The same applies to Kilimanjaro in Tanzania, Torres del Paine in Chile and many many others.

Now, you have asked about the nature of vortexes. Simply stated, vortexes are swirling eddies of electrical and magnetic energies. They are a function of the gravity and electromagnetic grids. They generally spin counterclockwise above the equator and clockwise below it. This is the natural energy pattern based on your Earth's polarity. Vortexes distribute energy outward in what are termed electrical vortexes and inward in what are termed magnetic vortexes. Some function as both. Mount Shasta, for example, is an inward and outward vortex-portal complex, Sedona is an outward vortex-portal complex.

Vortexes form for several reasons. The generally understood cause is the intersection of ley lines. We tell you they also naturally occur at points of tectonic stress, at volcanoes, around pinnacled and pyramidal mountains and around man made structures built to sacred geometry. Vortexes occur naturally at large mineral deposits, basalt beds, granitic batholiths, river confluences and water falls. You see, these all project subatomic plasma, charged ions, and electromagnetic fields. This natural build of energy will spin by nature, and thus the vortex occurs.

Now, we should clarify that vortexes are not portals, and they do not evolve into portals although from your perspective it could appear as such, because all portals will attract a vortex to them, you see. Vortexes can become the distribution engines for both adamantine essence and higher dimensional energies received through portals. Not all vortexes house portals, but all portals have a vortex. Do you see? The vortexes around mega portal complexes such as Lake Titicaca and its satellite Machu Picchu have double spin vortexes, spinning both clockwise and counter-clockwise. Indeed vortexes of this nature assist in the equilibrium of point zero energy fields and create antigravity fields that are capable of bending space time. Such vortex-portal complexes are potently drawing and distributing the higher dimensional energies of the ascension.

Now, one example of an enormous energy vortex that does not house a portal is Niagara Falls. The plasma created by the ions released from the massive crashing of water is extremely powerful and extremely beneficial to one's physical healing and emotional well being. A higher field of energy exists there, because the ratio of anions to cations in the plasmic field creates a very tangible sense of well being and elation. Niagara Falls is indeed a potent powernode. An enormous vortex complex spins there, but it is not a portal. This ratio of cation to anion is the same benevolent effect that makes ocean beaches with crashing waves so robustly pleasant and natural aquifer springs so inviting. Do you see?

Crystalline Fields

Those of you that have studied the crystalline structure of gems and crystallized metals have learned that the very atomic arrangement of gems and metals are geometric in form. Gemologists are well aware of this fact. Most diamonds form in molecular patterns of amazing octahedrons and dodecahedrons. We will add that not only do all uniform crystallized structures emit a complex electromagnetic field, they are also conscious and capable of many other utilities including increasing your life force through balanced amplification of your energy field.

Likewise, powernodes and sacred sites emit electromagnetic fields, because the concentrated energy matrix of sacred sites is frequencially crystalline in nature. As such they have properties of crystals, specifically the ability to receive, transmit and store light energy.

Like crystalline gems, each sacred site forms an elliptical energy field that draws within it the vital adamantine essence. Their field projection creates a geometrical lens capable of acting like an etheric prism, bending, refining and amplifying light in marvelous ways.

As a result your energy centers, your chakras, are both heightened and balanced. Thought waves are manifested into reality far more rapidly within these crystalline adamantine templates. The ancients understood this and thus utilized these places for prayer, meditation and vision quest. In such reverent process, a sense of well being is achieved and stored within the energy of the site. This sense of well being is a key part of why so many are attracted to and compelled to visit powerpoints and sacred sites. The profound spiritual links between Earth and the

Cosmos help you to better understand your great yearning and mission to know why you are here. They help you understand your purpose and activate you to evolve into a higher form of consciousness. Sacred sites are cosmic cathedrals that help you unlock the great mystery.

Humans tend to assign a singular chakric focus to sacred sites. One may say this place is a heart center, or that location is a crown center. In truth they activate all of the human centers. The focal center that a seeker may sense can and does depend to a large extent on the light quotient and needs of the individual visiting them. A sacred site which may be considered as a root chakra center may be genuinely experienced by others as a heart, crown or throat amplifier.

The channel asks if there are 12 major chakric centers for the Earth, and if each continent has a subset of chakric centers. There are chakras, sub chakras, powernodes and sub powernodes on every continent of the living Earth, but they are not necessarily arranged on that basis, although many of you classify them as such.

However, there are teams of 12 major planetary chakras. But you see, there are several sets of these, each teamed and aligned in unique purpose. Some sacred sites occur in all of these teams. So we clarify and note that a chakra for the living planet's life system is different from a chakric sacred site for humanity upon the Earth. Some of them are the same, but not all. You see, the chakras of the living sentient Earth do not relate to Gaia in the same manner as the human chakras would relate to the human bodies. In other words, one that relates to the divine purpose, well being and spiritual development of humankind may not relate in kind to the body of the Earth. Many of the chakras of the living Earth are inaccessible to mankind, being below the crust or below the seas. The Earth's spinning and pulsing inner core is the heart chakra for Gaia but not for mankind, you see.

Some of the 12 alignment sets are relative to the grid systems and to the ley line systems, others to the axial tonal and meridian lines of the Earth and to that which would be adamantine and etheric. Some are designed specifically for the living Earth, elements and directions to be able to communicate with humanity. These facilitate mankind's ability to communicate with the conscious awareness of the Earth. Others are specifically designed for humanity to communicate with higher celestial energies. It is partly because of this chakric system that we have this conversation now. Do you see? There are many types of chakric systems on the planet.

Ley Lines

Now, Dear Masters, we cannot truly discuss sacred sites and powernodes without the inclusion of what you term ley lines. It is a term coined to simply mean a straight line connecting two points, but our Metatronic term is much broader.

We define ley lines as being on the conscious crystalline aspect of the electromagnetic flow lines that network the planet. Ley lines are 'groomed' energy flows of electromagnetic energy currents that network the planet and, in a latitude of comparison, can be said to serve as the nervous system of Gaia. The raw Earth currents are called dragon lines. As both ley lines and dragon lines are electrical in nature, they flow along paths of natural electrical conductors.

This electricity occurs naturally on your Earth from multiple sources. Moving water such as waterfalls, rain and breaking waves produce charges as do the decomposition of organic matter, tectonic stress, volcanics, lightning, solar heat and winds. The very crust of the Earth's surface you live on with its electrically conducting gases, metals, semiconducting mineral crystals, water soaked organic matter and electrolytes offers an excellent medium for maintaining and producing electrical currents. The mineralogy of the strata below the surface conducts in kind. Charged ions are drawn to the ground, and high concentrations increase the intensity of Earth currents through the electrode effect.

As a result, static electricity flows parallel with the ground in an orthogonal quasi static electric field, forming eddies and springs of current that become rivers of current that network the planet. These streams of current flow along lines of conductive minerals such as iron, gold, copper and quartz bearing rock. Naturally they flow to mountains, volcanoes, waters, minerals and large deposits of other conductive formations and amass through, up and around powernodes. As you are aware, when these lines intersect, a spinning vortex is created.

This electrical current, as with all forms of conscious energy, can be directed. The first to do this were the scientists of Atlantis. The patterns of raw Earth currents, or dragon lines, were scientifically detected, mapped, and esoterically imbued and consciously willed to flow in specific routes. The scientists discovered that by flowing these energies through their Arcturian fire crystals and magnetic transducers that they could be amplified, refined and separated into a form of braided polarized crystoelectric frequencies capable of creating antigravity fields when spun in counter balanced vortexes. In the technology of the golden age of Atlantis, these conscious crystoelectrical lines became so technically refined that they were used to create a labyrinth of tunnels for both communication and transport, and they were crossed to form energy fields for varying purposes. They were patterned to flow inside factories to energize the work force and in agricultural fields to stimulate plant and crop growth by routing them in conduits relayed and amplified through crystal power stations, for these Earth energies were used to power homes.

The forefathers of the sect you term as the Druids were a branch of Atlantean scientist-priests. These were, in the golden age of Atlantis, learned men and women who understood the necessity of combining scientific law with the energy of the divine. This was termed the Law of Oneness.

They learned that certain of the crysto refined ley energies were capable of holding an intent and field of awareness. A global network of this category of ley energy was put into place, as the scientist-priests discovered how to use natural powernodes as relay and amplification stations for this energy to retain a self generating sufficiency. When these leys became integral with infinity apex points, the divine aspect was imbued within their energy awareness. These Atlantean priests, (called the Atla-Ra) with assistance from Acturian and Sirius B Masters, created specific divine ley energy routes connecting major chakric points on the planet. These could be programmed with sound frequencies and color to enhance the tranquility and well being of temples. They could be crossed to create sacred energy vortexes.

So in truth, there were many types of ley systems in Atlantis. Some were entirely used for crystoelectric power in various forms, others, the ones still operating albeit in fragments, were the leys of divine awareness.

One of the most robust survivors of this system are the leys now referred to as the Michael and Mary lines. Indeed, its level of intactness is due largely to the work of the Druids who had escaped Atlantis before the demise into existing monasteries in Britain, Europe, Egypt and Og. The most prolific were the sects in Britain and France, who used antigravity aspects of leys and sound to assist in the formation of stone circles. The Michael ley line survived for a myriad of reasons. It was amplified in stone circles and by the very Cathedrals built on its course to sacred geometry. It was not called the Michael line in its inception. Rather it was referred to as the Atlas Line in Atlantis and Thoth line in Egypt and Og. Its pagan name was changed to that of Michael and Mary by the secret societies of savants to protect it from the Church. The Freemasons, who built cathedrals capable of amplifying ley energy, utilized sacred geometry imminently. Almost all of the Cathedrals and Greek Monuments were built to phi, the golden ratio or mean, directly on powernodes along ley lines. Would it surprise you to know that Archangel Michael was very much present in the time of Atlantis but was called by a different name? You see, we show our love equally to humankind of all eras, races, religions and creed.

Now sadly, ley lines are not constant. In time, they shift and dive. And so what was once a robust system is now fragmented and shorn. The ley system currently is but a shadow of its former self and no longer completely circumnavigates the planet. However the raw current does, and many parts of it are indeed ley in nature. This system is currently under repair, particularly by those of Sirius B and by many of you in multidimensional aspects of yourselves. Do you see? Like powernodes groomed into sacred sites, dragon lines are quite capable of being imbued with energy from humankind and from sacred sites, and they become polished into ley lines.

Now, ley lines flow in specific manners, generally speaking. They will spiral counterclockwise up domed mountains above the equator and clockwise below it. They flow upward in straight lines on pyramidal shaped mountains. They spiral up conical peaks. This is why mountain peaks contain very high frequencies.

Preparation for Visiting a Sacred Site

Because the frequency within apex infinity points is at a much higher resonance than that of your normal dwelling, certain precautions are given. The energy differential of sacred sites to the energy of one's electromagnetic field can in time force equilibrium to occur. This can create fissure cracks in one's auric field, and an energetic bleeding can occur after one to five days. This can be helped by wearing a noble metal bracelet on each wrist and around the neck. Gold is recommended, but silver, brass or copper can suffice. We would recommend wearing a lapis lazuli of at least 10 grams about the neck as a pendant. The access to higher dimensions can also be facilitated through use of a true vogel crystal, cut to phi. Daily bathing in epsom salts, dead sea salts or mineral bath salts can also assist. Drink plenty of water, and avoid overeating. Infinity points are amplifiers, so be aware of your thoughts and emotions within these energies. Avoid feelings and thoughts of anxiety, anger and fear. What is in your mind and emotional field will be amplified whether it be thoughts of love or thoughts of a negative nature. Acknowledge the energy of place, and ask permission to enter and make an offering whether it be flowers, sage, tobacco bundles, a small crystal or cornmeal. Smudge before and after with white sage.

Finally, do not overstay your time until you are acclimatized. Those who live in areas such as Sedona and Mount Shasta City will in a matter of weeks or months reach equilibrium. But in the interim, emotional imbalances caused by fissure cracks in the auric field can and do occur.

The Quantum Crystalline Activation

Masters, in closing this discourse we leave you with the understanding that all of what we have discussed is around your work in achieving crystalline activation. You see, it is an essential acquisition for both seekers and for the planet. It is in fact occurring planetarily through myriad modalities, focally through the powerpoints, sacred sites and leys of the Earth. The vortex-portals of major powerpoints have crystalline geometric matrixes, very similar in nature to the octahedronal and dodecahedronal crystallography of precious gems. When higher dimensional light waves are received through these overlay lenses, the planet is imbued with platonic crystalline frequency.

Man through intent and impeccability achieves the crystalline vibration. What is impeccability? Impeccability is simply walking one's talk, living in integrity, eliminating fear and worry, always doing your best, honoring others and last but certainly not least achieving self love. Granted, this is easier said than none, but it is essential in activating your crystalline field.

Can mankind achieve activation of the crystalline qualities without visiting sacred sites? Yes, absolutely. But those of you who choose to do so will be within infinity points amidst a crystalline field that is already there, and in so doing the process is greatly facilitated.

You see, crystallization is not just an atomic molecular order in physical science and chemistry. It is also a vibration achieved through the metamorphis of crystalline alchemy, a vibration of integral clarity that refines the human electromagnetic energy field into the pristine sacred geometry of the Merkabah. When you accept to activate your crystalline field, you acknowledge, define and empower your Divinity of Self. You then carry the most precious of sacred sites – the human heart – with you everywhere you go. Within crystalline activation, one's heart emits a golden spiraled frequency from within the merkabic star tetrahedron, and this frequency sings the symphonic song of Ascension with the instruments of every sacred site orchestrated to crystalline perfection, and it resounds in ecstasy to the Cosmos.

I am Metatron, and you are Beloved.

… and so it is. AA Metatron via James Tyberonn

The Platinum Ray ~ Global Warming and Earth Changes

Archangel Metatron Channel via James Tyberonn

Greetings, Beloved Masters! I am Metatron, Master of Light. Now we send to you the 'Platinum Ray,' an energy of omnipotence, an energy of recreation. Draw it into your frequencies – the energy of platinum, of palladium – these are the energies of the ascension and of the new beginnings.

The Platinum Ray

The energy of the Platinum Ray is a multidimensional frequency having a unique utility that is manifested and drawn into the Ascension phase of planetary systems and humankind. In truth Dear Ones, it has been used by many of you in other dimensions and planetary systems, and many of you chose to be here at this time in part to reexperience what you already know and to help others in their experience of the Ascension. Indeed the Platinum Ray is the ray of crystalline light, the Metatronic Ray of the Earth's Ascension. Draw it into your hearts as you face the quickening of this marvelous event. It is the most pristine vibratory ray in your solar system at this time.

The Platinum Ray is a soft frequency often seeming feminine in nature, but in truth it represents the optimal balance of both male and female. Indeed it is the balance that is required for the new Earth. If it were either male or female in resonance, it would not have the equiponderant harmony that is requisite in the new paradigm. Indeed it is the energy of the Cetacean, the dolphin and whale of Sirius A, that joyful ecstasy of the platinum silver realm. Engage its velveteen tranquility and encapsulate your hearts with it as you move into the heralded Ascension. Engaging the Platinum Ray will empower you, Dear Ones, to enter the coming changes in composure and argentine serenity.

Coming Changes

You are entering the threshold of a 'Cosmic Leap' and enormous graduation, and it involves many aspects. It encompasses many tools, many components, copious events that may or may not resonate with your sense of the ascension. One of these is Global Warming, and another is Earth Changes; the two are sequentially twinned. Most of humankind fears them. Be aware that such fear is only the reaction of your conscious filters within the plane of duality. The greater part of you celebrates this destined event.

Global Warming and Hurricanes

The events of increased hurricanes, increased tornadoes, solar storms, earthquakes and volcanic activity are ALL tools of the Ascension. You see, each increases the terrestrial frequency. Each is a mechanism of Global Warming and as such of Earth Changes.

Now, the channel asks about the increase in hurricanes. In truth, the hurricanes are an extremely robust and vital accoutrement of energy distributions intended for the oceans. The amount of energy in a cubic meter of a class five hurricane is as great as an atomic bomb. These are intended to raise the resonant frequency of the oceans and are not meant for land. In truth humankind, particularly in your United States, draws the energy masses onto shores by the very focus blasted across the media to millions of attentive viewers predicting where they will make landfall. Is it this city or this county and how has it progressed. It becomes almost as a game and hooks the attentive fear of millions, and that very mass of energized fear pulls it onto land.

In truth this quantum surge is an event of sacredness, a grand occurrence to be celebrated. Yet mankind may not all see it as so. You see, it involves the forementioned Earth Changes and the current mechanism of global warming. In Cosmic terms, in Cosmic chronology such an occurrence of graduation is not so rare. You see, it is a necessary progression, and in celestial terms it may be considered a planetary generative pubescence. It is an event that all of the celestial bodies in your solar system have already experienced, and Dear Ones, many of you were on those planets when they ascended, as all of you exist in multidimensionality.

Now, there is much sibilation in your media and among your academics on the advent and causes of global warming. Indeed there are even arguments as to whether or not it exists. There is finger pointing and accusation and denial. There are theories on the cause of global warming, discussions of greenhouse gases, of C02 emissions from fossil fuels and recommendations to change it.

Masters, the primary cause of global warming is not greenhouse gases, and it is not caused by burning fossil fuels. So although many of you are eager to take the responsibility for planetary warming, humankind and pollutions are not the drivers of global warming. Does that surprise you? It is true, so do not be so eager to berate yourself, Dear Ones!

Now, we do not attempt to chastise or disagree en masse with Al Gore whom we tell you is a soul of great heart, the reincarnation of your President Lincoln, yes the same! We agree with him in intent and say it is indeed wise to conserve, wise to find alternative fuels, wise to stop pollution. But these activities, while they have contributed in a small degree to acceleration of the process, are NOT the prime drivers of global warming. Global warming indeed would be occurring even if you were not an industrial fossil fuel driven culture.

The principle cause of global warming is the increased spin velocity of the planetary inner core. It comes from the very center of the Earth. The changing torque, the increased pulses that flow outward contribute to the melting of the caps and the changing of air and water currents.

The warming of the planet is indeed occurring, and it is part of the Ascension process. Fear it not! Nothing will stop it, nothing should stop it. The ice will melt, the waters will rise, the hurricanes will continue. That is a strong statement. Yet it does not stand in defiance of the fact that your future is a moving target. You see, global warming is not a probable reality, it is not a prediction, it is a certainty, and it is occurring in the NOW. That does not mean that catastrophic change will occur. That is yet to be determined.

Earthquakes will continue to increase, and the spinning and tilting will continue. These phenomena create aquatic quakes, tsunamis, and this will particularly occur along the Pacific ring of fire. So the increased spin velocity within the planetary core has created an intensity of friction between the mantle, the crust and the magma of the planet. The increased velocity of the core has created a greater heat in the molten iron nickel composing the spinning core. The greater heat of the core physically yields a greater core volume, you see ... physics 101, yes? So the pressure of the larger mass volume of the core, its expansion's friction, has produced a consequential heating of the planetary surface, oceans and jet streams. All quite independent of 'greenhouse gases,' we will add.

So indeed, the expansion and contraction of the Earth's mantle will create and is creating a friction that is heating the globe. It will create eddies in the airs and waters that will induce winds, hurricanes, tornadoes and such, and it will continue to do so.

Ocean waters will rise, and island nations will disappear. In other oceans, tectonic forces will push new lands to the surface. Rivers will change their courses.

Many of you will recognize the patterns of the coming floods; many of you will recall it almost on a biblical prophetic proportion, an Atlantean memory. Great truths will come forward. Great wisdom will be attained. This will unfold, this will continue ... but it is not as yet certain how it will unfold. It does not necessarily need to be a global cataclysm. It does not need to be catastrophic. It could be more on a regional basis and a gradual shift over two centuries in time. Much depends on what we term the light quotient, the light quotient of consciousness.

Now, fear itself will bring forth many experts, many who wish to speak. Many seers will come forward and predict exactly how and exactly when this will unfold. Heed them not, for although they may not all be mal intended, it is yet to be fully known how it will manifest; it is yet to be determined how and to what detail of intensity the changes will unfold. You see, when channels make a time based prediction, they do so from a specific point in time and space, while the potential is in constant flux, ever changing as it moves forward. That is why it is truly not the purpose or nature of the divine to predict time line events. Be aware that is not our purpose here; we speak of the trend of the ascension, and we tell you that the ascension, the warming and changes will occur, but the timing and details of such are YET to be known. In that detail nothing is written in stone. Use discernment, and Masters be wary of those seers and channels that focus on prediction tied to time and seek credibility and notoriety for same. That is not the aim of Spirit.

Dear Ones, regarding the Earth changes, there are sundry overt and enigmatic perspectives and nuances occurring here, each presenting multitudinal potentials. The Earth changes speak to the soul, they speak to the future and to the past. They speak to consciousness, to light.

The warming and the resulting terrestrial changes are an intricate activation, a potent vivification of planetary consciousness. And as such, the melting of the poles reveals an entanglement of great complexity. The polar icecaps hold a far more complex purpose on the planet than mankind has understood. They are crystal, you see, crystalline ice, and these have recorded and projected the magnetic and frequencial codes of the planet for millennia.

Not all humans will chose to remain on the Earth during this transition. At this time much will be revealed about existing space stations that are shielded from the masses.

Many will say this cannot have purpose, this cannot be an act of ascension, this great change, this great exodus ... yet it is just that. It will be incumbent upon you humans to find your balance, your well being and your own sense of peace throughout the evolution of the planet, whether the experience be one of relative ease or one of great turmoil.

And we will add the caveat that these changes and their degree of hardship will evolve according to the quotient of light. It need not be as catalytic as the previous one, and the current probability is that it will not. We repeat, the current probability is that the poles will shift, the seas will rise and the Earth will change in a manner far smoother than the trauma that you saw in the sad demise of Atlantis. You see, the light quotient among humankind on the planet is at a greater frequency and at a higher percentage in your current times than those of 12,000 years ago. The current probability is that the changes will occur without the devastation of total global catastrophe.

Nonetheless, the probability of the current course suggests that there will be more regional cataclysm on the scale of that which was witnessed in the tsunamis of Indonesia and eastern Asia. Many will choose to exit. Those that do remain will relearn how to create their sustenance from consciousness.

You have been told that many of you waited in line to be a part of this heralded Ascension experience. In reality there was no queue as if you were lined to enter a theatre, but there was and is the great and magnificent desire for this experience, to be a part of this completion. You see, ascension of a celestial body is a divine event, an even greater event than the birth of an avatar of a Cristos onto your Earth Plane. The ascension is the event your prophets called the Second Coming. And, Dear Ones, to be a part of it is a credential imprinted upon your soul, a badge of honor.

You see, there will be a time in the distant future when people will ask was there really an ascension, was it a wondrous magical time, was it the birthing of a new energy.

So of course, Ascension cannot occur without change. The increased spinning of the Earth's molten core is the root cause of the melting of the ice caps, and this must occur. Within the ice crystals of the poles are the frequencial codes of the planetary resonance, you see. The crystals, the ice crystals project an enormous code and enormous electromagnetic field according to the crystalline structure, and therefore they must be dissolved, must melt in order to recreate the planetary resonance, you see.

So Earth Changes are a necessary part of the transition that will take the planet beyond the 3rd dimension and into the 4th, 5th and beyond. Consider it a rebooting in order for a new program to be installed, the program that will allow for greater dimensional access in correlation with the 144 Crystalline grid forming around the planet. Do you see how it all ties together?

Earth transitions are being monitored on many levels, both as an upcoming event and as a paradoxical probability. Many of you, including the channel, are aspects of much larger beings who are fully aware of the roles you play on the inner planes and upon worlds beyond this one. As we have told you before, many of you have what can be described as aspectual parallel existence in the realms of the Masters from the Pleiades, Sirius B, Arcturus and from what you would term, from your earthen perspective, your own future ... although in multidimensionality it would not be seen as such.

Now, the issue of the Earth's magnetic poles and the Earth's axial tilt are intricate to the Ascension as well. These are essentially two different yet related phenomena. One is the reversal of the planet's magnetic poles, and the other is a full geographical tilt of the Earth on its axis. Both have occurred at various times in your planet's evolution. The probability of the current flow is that one but not necessarily both will occur, and that is the reversal of the magnetic poles. In fact, it is already partially underway. Your scientists, in fact, have considerable evidence of this occurring already. This event can occur without cataclysm, can occur internally, and the degree of its influence is yet to be determined. But polar reversals occur regularly on your sun and on your moon and indeed on your Earth albeit to a much greater time sequence, and these can occur, relatively speaking, rather smoothly. Ah, but axial tilts are another case, and an axial tilt would indeed have much stronger effects, you see.

Now, the melting of the polar ice caps will indeed influence the weight distribution of the planet, and the shifting of such density must be accounted for, must be compensated in centrifugal balance. And as the polar caps must 'reboot,' we tell you the slower process of Global warming is by far the more gentle method of this reprogramming in comparison to an axial tilt. Mechanically speaking, the axial tilts of celestial bodies are primarily created by an unbalanced proportion or shifting of weight upon the planet, by rapid magnetic shifts or by impacts from asteroids.

Keep in mind that solid mineral and rock are far more dense than water. A cursory glance at the globe of your Earth will show that the continents are far from evenly distributed. On one side of the planet is the enormous landmass of Africa and Europe, and almost immediately opposite this there is the wide open expanse of the Pacific Ocean. It is in part why your planet has a 23.8 degree tilt in its axis at the present time. The Earth's greatest north-south expanse of landmass, from the tip of Norway to the tip of South Africa, is directly opposite to the world's greatest north-south expanse of unbroken ocean. Quite a disparity of weight, but it does not necessitate an axial tilt ... this can be compensated by magma distribution.

You see, Ascensions are not mere mechanical events, they are consciousness shifts, they are divine in source ... they are not totally ruled by physics. They are purpose driven and occur in pristine unfoldment in keen awareness of that which you call divine. Do not, Dear Ones, omit the sacred, the will of the divine universe.

Such shifts have a cyclic aspect. They occur within parameters of cycles determined by light quotient. We have told you that space is congealed time, and time is liquid space ... and with your acceptance of that postulate we tell you that your Earth operates on a cycle of time. We tell you that the cycle of the Earth is that of the 12, and that approximately every 12,000 years, taken from your current dimensional perspective of time space sequence, the Earth completes a cycle, just as your human biology does every 7 years.

However, even a relatively non catastrophic magnetic reversal of the poles will result in a lessening of the planet's magnetic field, allowing damaging solar rays to penetrate to the surface of the planet. Such an occurrence will necessitate the vibratory helixed energies of platinum and gold to provide a filter of such energies, among many other functions of what will be termed the Crystalline Firmament: a protective field of the platinum ray that will emerge after the completion of the 144 Crystalline grid matrix.

Dear Ones, we leave you with a message of joy, of strength and of great honor. What you are doing requires great integrity and fortitude. You are experiencing a crucible that will purify and engineer a world, an age, an Ascension. Do you know how much you are held in esteem and loved for the courage of what you are doing?

Veneers of negativity, fear bases will be set aside, as power in pristine expression is redefined, reclaimed and aligned in its highest frequencial format. The appropriate use of vital will, of strength, of power will be tantamount in the Ascension. You see, when strength is virtuously embraced and expressed purposefully, it will be rewarded with the exertion of peaceful will, which is power's true companion, its intended equal. As such, external expressions of peaceful intervention through intent will be ubiquitously appropriated, potently projected by those who have achieved, frequencially pure vitality within their own experience, molded by the resonance and process of the ascension participation.

Dear Ones, much awaits you, and all of this is molded and assisted by the Platinum Ray, the ray of Ascension, the ray of rebirth, the ray of purity in creation, the Ray of Metatronic change. It is a helix of platinum and crystalline gold. Now, among the tools of projection of the Platinum Ray are the metals of platinum, palladium and rhodium, as well as pristine quartz created in the Vogel Cut to phi, those of the facet 44 and 88.

You are awakening to the graduation and perception of Crystalline Light. Crystalline Light precedes the Ascension; it is faster than light as you know it. And from where does it come: from that of the Gamma Burst, from that of Enoch, from that which you term the 'Black Holes.' So much awaits you, Dear Masters.

I am Metatron, and you are Beloved.

… and so it is. AA Metatron via James Tyberonn

The Solitude and Sovereignty of the Divine Self

Archangel Metatron Channel via James Tyberonn

Greetings, Beloved: I am Metatron Lord of Light, and we welcome this gathering!

Now, the channel has asked that we speak on Solitude. He queries whether solitude or partnerships bring greater advancement to the soul. The question is timely and especially pertinent for many of you in the dawn of the Ascension.

The short answer is that both coupling and solitude have their divine purpose ... and we underline the word BOTH. Much is gained from being in committed loving relationship. This is the natural circumstance during most sojourns on the plane of Earth.

But Masters, we tell you that there is also great purpose in solitude. You see, a great many of you find yourselves alone during this period, this prelude to the coming Ascension. If you are among these, we tell you that perhaps your solitude is appropriate. Perhaps it is as it should be. You see, it is how many of you planned it. Now, many of you feel that the loneliness is too much to bear and that you must pursue a partner ... the elusive twin flame. Dear Souls, there is much confusion around the concept of the twin flame, the 'soul mate' and the role and nature of optimal partnership in Mastery of Self.

Now, there is a recurring message in many of your religious texts that says, "For everything there is a season, a time for every purpose under heaven." Indeed there is a time for solitude. Divine Solitude is a noble condition and one that offers quantum leaps in terms of growth, when understood and recognized for its profound purpose. You see, the Divine Self is ultimately conclusively alone in its final pursuit of omnipotence. In your terms, the soul enters the Earth alone, and the spirit departs the Earth alone. Ascended Masters that walk the Earth have for millennia sought solitude in their incarnations before achieving their Mastery. So it is with many of you now.

The present phase is a uniquely opportune juncture for self review and chosen solitude. Masters, many of you particularly those above the age of 49, the seventh 7-year cycle, find yourselves in a state of solitude after spouses have passed, relationships have terminated and marriage contracts have ended. So we tell you to use this time wisely, embrace it. You are on the cusp of a great graduation. Solitude avails ones' self for those seeking sanctified light, for those seeking what is termed 'consecrated enlightenment,' choosing specific periods in their life plan to be alone for a time. This does not mean you will always be in solitude, it simply means that for a time you have chosen it to 'work on the self,' to achieve self love.

Love of self is a condition that many, especially those from Christian heritage, have lost. Christianity with its teachings of original sin taught you that you were flawed in your nature. You spent lifetimes prostate, asking forgiveness for who you are. You lost your sense of your divinity and found it easier to give than to receive. Balance was forfeited.

Solitude is intended to be a period of sublime reverence of self. Your life and your experience in this plane is your own creation, your own living tapestry woven by your individual belief. Within solitude the soul is prompted into self-review, and opportunity is given to dive deeply into the deep waters that flow within you. To swim in the ocean of SELF and in so doing rediscover the love within, to learn what a brilliant spark of God you truly are.

Relationships are a method of reflecting the affectivity of your belief system and of giving you feedback on what, simply stated, is working and what is not. Detachment requires the individual to explore the self, to reacquaint with the inner horizon, and this facilitates and necessitates sovereignty. Sovereignty is the pre-potency of Mastership. A relationship of two sovereign humans has greater balance, greater creativity and greater longevity than a pairing of two beings dependent on one another. Do you understand?

Ultimately each soul must clearly define SELF in order to gain Mastery. Self Mastery is embodied in periods of planned detachment. It is that period in which impeccability is crystallized. And we tell you, Dear Ones, crystallization through impeccability is a necessary phase of Self Mastery. It is a calibrating juncture in the multidimensional sojourn. One enters the void, the great mystery in the quest for fortitude and sovereign vision alone, without a shoulder to lean on.

Do not misunderstand our meaning; there is great validity in coupling, in the natural aspect of twin flame and soul mate. But conclusively one walks the path of Mastery in detachment.

One becomes enlightened when one learns to transcend the physical self. Each of you must endeavor toward the final conquest of what we term as 'impeccability.' Impeccability is the crystallization or uniform clarity of the soul, and it is a necessary virtue of Mastery. This involves release of dependency, the release of all that does not serve your divinity. It is a rebooting and reprogramming of all you are.

We have told you that the language, the fabric of higher dimension is sacred geometry. Impeccability is the geometric clarity of the soul mind. By defining oneself through impeccability, one becomes crystalline and thus more capable of consciousness within the geometry of higher planes. It is only accomplished by deciding who you are, what you believe, and then living it recognizing your truths and aligning fully to them.

Question to Metatron: Are you saying that loving relationships such as marriage are not our ultimate union?

AA Metatron: *In a manner of speaking, yes. Relationships are a means to the end. We are saying that loving relationships are a sacred, joyful tool of achieving Self Mastery, but that ultimately in each soul's journey there is the growth into sovereignty. The sovereign self is a sufficient self and truly has no dependent need of another. Such conceptual dependency can be a deterrent to Mastery. The paradox is that a relationship of two sovereign beings has greater joy, greater balance, greater interface with the divine than a relationship based on dependence. Do you see?*

So we say embrace your chosen period of SOLITUDE; it is the sign of your soul's intent into Mastery. We add that many who are in this life of the female gender have chosen particularly efficacious roles in the balancing of the planet at this time. Is it not true that the planet has been imbalanced in an overage of patriarchal energies for millennia? That is why it would appear that a vast majority of those drawn to the 'New Age' are female, you see, to anchor in the Divine Feminine. Females have been conditioned and labeled in your current paradigm as the 'weaker sex.' Rubbish! Nothing could be farther from the truth! Many of the females who find themselves now in single status are conditioned to feel they must have a partner. Again we say, embrace your solitude. You chose it; we honor you for the path you have chosen.

The ideal for the planet is not to be female or male rather a perfect balance of BOTH, but at the moment it is still in an *imbalance of patriarchal resonance. We honor those of you in female biology; your strength is indeed progressing in creating the nurturing balance. The perfection of paradox, the beautiful balance is that many of those in feminine biology in this sojourn that lead the charge for equity were in male bodies in their recent past and helped to create the very imbalance they now seek to rectify. There is symmetry and beauty in this trut***h.**

Truly the soul is androgynous, self contained, self sufficient. That does not mean that love is not the frequencial basis of the soul, indeed it is the highest vibration. It is the resonance that is produced by the Ascended Soul to the Cosmos and reflected back as a collective harmonic of crystalline love. There is a time when those of you who have played the role of soul mates will individualize and, in your terms, part in joy. The evolved soul, in achieving omnipotence, will become consummately self sufficient and in so doing radiate spectacular unconditional love to ALL. That is as it should be, as it must be. It is how you Ascend.

The final graduation of the soul is not done in pairing, you see. It is done within the Divine SELF in universal harmonic to the All that is. Do you understand? This is the activation of Self to the divine resonance of the Quantum Crystalline Field. Separation must osmotically occur before the final collective reunion; it must occur to allow for the final coalescence into the ALL THAT IS. It is who you ARE. It is the I AM that I AM.

Question to Metatron: In current times a lot of marriages seem to end after varying numbers of years, and the life long commitment does not seem to endure. Has that paradigm changed?

AA Metatron: *Ah, we enter into a very delicate topic, one often misunderstood and one that has many nuances. One's family* *is truly composed of all souls, all mankind. And in the lesson of Earth, of duality many forms of relationship, of soul 'contracts' exist, do they not?*

Now, we will say that 'lifetime' marriage relationships in the current times truly do not work within the new paradigm by the rules of their current setup and legal concept. Some persevere, most do not. We will say that it is better to end a relationship in which the creative flow of love is blocked than to 'wait it out' in dysfunction, contempt and drudgery. Many souls who end a dysfunctional relationship find that the relationship actually improves once the 'marriage' is ended. Is that not so?

We will add that the current level in which marriage relationships are ending in divorce is at an all time high on the Earth plane. This is global, not just in America and Europe. Additionally the degree of intimate relationships of cohabitation that end without consummating a 'legal marriage' is also at an all time high. You see, the model is changing. Partners will continue to choose different partners and in some cases different genders, and this is also a part of the new paradigm.

Balance in relationships is essential, and partnership 'contracts' are as varied as the flowers of the field – varied in form, diverse in color, in gender, in purpose, pattern and duration. Yet all are compelling. Love has many forms. The human DNA is changing. DNA is in 12 strand helix, and it is indeed the divine interface of human biology. The full spectrum DNA is in fact the catalyst of the changing prototype of relationships, and these are evolving as they must. Relationships are chosen for purpose, and the archetype of their varied criteria has changed and will continue to change in quickening times. Change is the nature of reality. Some of life's greatest teachings are garnered in the review of relationships that end as well as the review of those that bloom.

Question to Metatron: What about rearing children and the effect on children from broken marriages?

AA Metatron: *This is also a question and issue steeped in the complexity of change, of culture and of your belief systems. There were phases of human existence in Mu, Rama and Atlantis where children were raised in commune status. It was the accepted cultural methodology of child rearing and in certain ways was superior to what has occurred since. Children indeed knew who their biological parents were but were afforded a group upbringing that allowed each of them equal status, equal love, education, opportunity and health and emotional care from devoted loving caregivers. That seems quite foreign and somewhat cold to most of you in the current space-time sequence, yet the system had many advantages to that which occurs now. In current times on a global basis there are literally millions of children who are essentially abandoned, functionally orphaned and have to fend for themselves in not only what you term the third world but also in America and Europe.*

The fact is that many of the children today in the 'first world' are children of divorce and in your terms from 'broken marriages.' These children clearly understand that the current system of marriage simply did not work. As these children grow into adulthood, a new relationship prototype will emerge. It will emerge into a new legal structure.

A new system of defining relationships, of legalizing marriages will soon emerge. These will be structured on legal agreements of a shorter term. There will be the 'Five Year Marriage Contract' available in some countries, and these will be much more suited to the reality of the current relationship structure. These marriages are in essence legal agreements of shorter term, clearly contractually defined and will end according to the time limit or be optionally mutually renewable at the end of each phase, with room for adjustment of terms. Clauses will contain conditions that include welfare of children and define the appropriate responsibilities. These will grow in popularity and will spread globally as the norm. These will work. This trend will soon begin in Europe and spread quickly and widely. The new marriage contract will of course bring its own challenges and responsibilities. They will afford their own promises, promises of a new paradigm, but there will be avenues and opportunities of adjustments and the ability of reviewing terms. This will allow for greater recognition of the SELF. These will be designed to prevent energy blockage and reduce dysfunctional marriages, you see. Some relationships in these terms will indeed last a lifetime, most will not. But the archetype will better support both, based on choice and not on imposed terms of morality or cultural pseudo religious forces. People will see themselves in this framework as being 'whole' and not being their partner's property or someone's 'better half.' Each will bring their integral whole SELF in fullness, in robust flow of energy into relationship in a manner that often does not occur today. There are pre agreed exit points that allow for freedom of choice and for change. People will share their best without compromise of individuality. Each will recognize the divinity of the other and retain the dignity of SELF.

Religion

Religion will also undergo further change in the aftermath of the Ascension. Of course the established control-fear religions particularly Christianity, Islam and Judaism will resist. But in the end, their inability to maintain and attract followers will force them to change, dissolve and reemerge in less restrictive formats with more positive messages. Religions that control through fear are already losing their grip, and this will continue.

Humankind in mass today truly do not recognize or understand their soul, their divine SELF. Self is unfortunately relegated to the realm of ego personality by the masses. It is true that a higher degree of the light quotient is awakened on the planet now than has occurred at any other time, but it is still only about 10 percent of the population of six billion plus that has awakened or is on a probable path to awakening. That is a sufficient number to bring about the Ascension, but much has to be done.

The aspect of mankind's sojourn in multidimensionality is truly not understood, even by those who are aware of the reality of what you term as reincarnation. Western religion does not teach it, and the Eastern religions that do are not fully accurate. Lifetimes are not linear. They all occur at once, in what can be termed the quantum field potential.

The aspect you refer to as karma is also misunderstood. Karma is not a means of punishment. That which you term as 'sin' is in truth 'knowledge not utilized.' Karma then would be an action of the self, the higher self to enable greater soul learning, greater soul growth, greater soul awareness in order to achieve BALANCE, achieve self balance with the multidimensional self. All 'karma' occurs to and from the SELF; it is not owed to another.

Man's 'inhumanity to man' is one example of this principle. Less than two centuries ago there existed great brutality in the form of slavery. There has been gross iniquity and savagery inflicted on the indigenous populations in the Americas as well as barbarous ethnic genocide. We tell you that some of the people of the black race that fought to establish equality and 'civil rights' in the Americas in the turbulent 1960s and 1970s are in fact re-incarnated slave owners. Many of those that now find themselves trying to salvage the scattered culture and lands of the Native Americans are reincarnated government officials, gold miners and conquistadors that created the very conditions they now live in, again drawn to experience the counter side, the imbalance of what they themselves helped to create. Do you see? Part of the reason that so many casinos are popping up on Native American lands is because the ones organizing this are reincarnated gold miners that once ravaged the lands and people in plundering out the gold. They are now seeking a balance albeit in a rudimentary manner by attempting to bring it back in. Not necessarily the highest of frequencies yet the pattern is clear, and it indeed fits the bill within the karmic level of such souls.

So karma is not a punishment; it is the desire of the self to find balance in the multidimensional aspect of the soul. This is largely unrecognized and certainly not taught in any religion, yet it is indeed the truth, indeed the reality. Truly the soul reincarnates into all races, all genders in the eternal NOW. Do you see the beauty of the balance requisite chosen by the SELF?

Religion in the new paradigm must be individualized, must truly recognize the nature of SELF, the nature of the DIVINE SOUL. None of your world religions truly answer these questions today. None of your religions answer the questions of man's true Cosmic, extraterrestrial and multidimensional origins. And that must come to understanding in the new paradigm. None of your current mainstream religions can accurately and completely express the true history of man on the Earth. As such, there is no one true religion on the planet today. Most are bought and sold on preset regulated templates. Each claim to know the path to God, to be able to lead the way to God yet none truly do. Each has its dogma, each has its hierarchy and controls and each requires monetary donation and financial support from its followers. Dear Ones, Creator God is not short of cash! But your churches certainly seem to be.

None of the current religious dogmas are completely correct. The religion closest could be said to be that of the Tibetan Buddhist, but it also has its shortfalls. Some who have followed the path of the indigenous shamanism have found their way, but it is through individual perseverance and intent. Through solitude!

Man searches for GOD more fervently now than at any other time on the planet, and so this seeking in itself has the potential to bring in the light. Few religious teachers are true teachers, scholars perhaps but not true teachers, you see. Even among those in the so-called 'New Age,' those that channel are subject to the frailty of ego and the frontal mind filter of personal systems of beliefs. So discernment is the key.

Be wary of New Age Gurus that over pursue followers. Many get so tripped by their own celebrity that they sadly become energy feeders off the admirers that they draw. It is not only the southern evangelical leaders that are in it for the money. It is everywhere, even in the 'New Age.' Ego and greed are always there, always seeking the opening, and many rise but to fall. Spiritual one-upmanship is rampant.

When integrity is not maintained the connection to true spirit is disconnected, and all who are human are subject to fall in and out of integrity. Truly the way in the Ascension is the ability of each soul to rise into his/her higher self. Look inside and find your own divinity within your heart, not through blindly following a guru, channel or spiritual leader but through SELF. Accept only what you discern individually to resonate as true,

Dear Ones, and do not give your power to another. Each of you can and must channel your higher selves. Study, look, listen, discern, review and only accept what resonates within you.

The way to the divine, Masters, is through the sacred sovereignty of SELF. And in the Ascension the way to the higher SELF is through self definition and through seeking that aspect of God inside each of you with the great desire that is embedded in each of your souls. Study, seek and work! There is little hope for the lazy. You are here to make known the unknown! Work at it! Be a warrior of light, for the path is not easy. But within it lies an energy so exquisite that it fuels all that you require to move forward, and there is another level above each that you ascend to. Consciousness ever expands.

Discover your path. Discover yourself. Love yourself, love one another, love the divinity inside you and inside everyone! Discover the multidimensional aspects of your true Soul. That is the nobility of solitude.

You are Beloved. I am Metatron, Lord of Light.

… and so it is. AA Metatron via James Tyberonn

The Alchemy of Phi Crystals

Archangel Metatron Channel via James Tyberonn

"The crystal is a neutral object whose inner structure exhibits a crystalline state of perfection and balance. When it is precisely cut to the proper geometric form and when the human mind enters into relationship with its structural perfection in the vibration of love, the crystal emits a vibration which extends and amplifies the power and grasp of the user's mind. Like a laser, it radiates energy in a coherent, highly concentrated form, and this extremely benevolent energy may be transmitted into objects or people at will. This I know to be true."

Dr Marcel Vogel

Greetings! I am Metatron, Lord of Light!

We once and always greet you warmly and lovingly in this eternal Moment of Now. Indeed, we sense each of you personally at the specific time in which your eyes read these words.

And so another moment brings us together, uniting truth, understanding and wisdom through the matrix of the unified field, combining geometric thought pattern with desire and awareness. And so do all things come together in the same way. For all thoughts, all things, all beings are electromagnetic representations of consciousness, of manifestation, of all that ever was or ever will be. And so some things, some thoughts, some creations travel at more rapid speeds, more accelerated frequencies within the matrix. Others are denser and travel more slowly, more sluggishly. But all the same, they seem to arrive in perfect order, aligned with the nature of their creation and creatorship, you see. For the Universe, the Cosmos is perfection itself, regardless of any dimensional interpretation or perception of it. In a mechanical sense, in quantum tangent all things in the unified field act precisely, perfectly according to their nature, their architectural integrity. Yet some are created at higher frequencies, at higher vibratory resonance than others. And so we continue speaking about the Crystalline Field, a field that is of tantamount importance to the planet as the Ascension draws near.

Question to Metatron: Can you speak about Vogel Crystals, phi cut crystals and the man who invented them, Marcel Vogel?

AA Metatron: *Indeed! The inventor of these of course was an astute Crystal Master from the Atla-Ra, the scientist priests of Atlantis. However, he did not truly invent them, rather did he channel the knowledge of same which coexists within multidimensionality. In truth, most of that which you consider to be invented is not new technology, rather is it pulled from the future and from the advanced technical knowledge of the past.*

Crystal knowledge is essential to the coming age of the Quantum Crystalline Field. Many of you were involved intricately in this lost science. It is why so many are now fascinated with not only the quartz variety but all aspects of the mineral kingdom. Yet, be aware that the Crystalline Field is equally valid in terms of its relationship to consciousness, to thought, to action and to manifesting creation.

Marcel Vogel correctly surmised that the energy signature of quartz was very similar to that of pure water, and that as the human body is over 70 percent water the quartz could be used to charge and influence both.

Crystals cut into sacred Phi angles are capable of many things, most of which are beyond the current understanding of the masses of humankind within polarity consciousness. You see, these crystals exist above the spectrum of your visible light, well into the perceptual divisions of crystalline light within X-ray and ultraviolet. In so doing they exist in the eternal void or Now, beyond time sequence. As such they are time gates and can offer glimpses of realities both in your perceived past and future. As simply as I can explain it, your consciousness within planetary duality and 'flashing' continuum sequence is 'divided' into linear time and probability parallel areas. The Now moment in the higher ultraviolet field is a crystalline unit which contains Universal Truth. When you are able to access that unit, you achieve the ability to expand that truth and gain the control of your reality. Phi Crystals then are the fabled Philosopher's Stone that assist your consciousness into the unit of the Crystalline Alchemical Moment of Now.

Alchemy has always made extensive use of analogy, symbolism and so forth to relate transformational concepts to esoteric ones. In some epochs and contexts, these metaphysical aspects came to predominate, and the myriad transformational processes were then viewed as symbols of spiritual processes.

In your hermetic view of alchemy, the Philosopher's Stone represented the tangible crystallization and morph transmutation of a subtle substance into something greater. This then is the sacred metaphor for the inner potential of the spirit and reason to evolve from a lower state of imperfection into a higher state of enlightenment and perfection. Phi Crystals are the tool of the MerKiVah that takes one to the Alchemy Journey. They are one of the paths of the Philosopher's Stone, the crystalline journey into coherent energy through Phi.

The 'Black Hole' Within Quartz

Now, there are some interesting aspects related to certain crystalline forms that are only beginning to be understood by your scientists. Marcel Vogel correctly theorized that there were 'missing' silicon atoms in the quartz molecule (SiO_2), in what is now termed the 'Lattice Defect Theory.' There are in fact 'holes' that occur in the tetrahedronal pyramid of the quartz molecule. The missing component is silicon, and we tell you that space retains an energetic form that is capable of receiving and amplifying human intent. It is a 'black hole' of sorts that is a void space capable of receiving human energetic intent! That is why quartz and many other crystalline gems are capable of being 'programmed', and Dear ones, this is an important aspect of why these are particularly beneficial to you. Phi Crystals by their sacred geometric construct emit a unique energy, a crystalline light. They change the vibration that affects the able user. They change the frequencies at which humanity's minds operate. Phi Crystals receive human intent, and indeed intent creates an energy that is synergized and transformed through crystalline light. And that embellished energy is capable of manifestation and reaches the heavens and earth, the inner and outer dimensions, you see?

The variations of Phi Crystals are compatible with humanity. Some of them are compatible with the Earth, some with the elemental and emotional worlds and others with the celestial realms. Some create frequencies that unlock ancient encodements even as some natural crystals do. Certain Phis have past and future sequencing keys that open time and dimensional gates, but these are less understood in the present, and many of these remain unutilized and locked. This knowledge is now available to the savant and focused adept. It will surface and offer itself as the individual user becomes capable of understanding the responsible usage of same.

Now, your gemologists and crystallographers are aware that certain crystalline gems including topaz, tourmaline and especially quartz have a quality called piezoelectricity. This is the quite unique ability within the living crystal to generate electricity under specific polarized conditions.

Crystalline solids have a very regular atomic structure: that is, the local positions of atoms with respect to each other are repeated at the atomic scale. These arrangements of crystal structures serve many purposes and in essence form the life force structure of the Crystal Entity. As with all life there is an auric field and indeed a consciousness within crystals. That of quartz offers many functionalities that make it especially beneficial to mankind, yet in truth all crystalline structures offer myriad benefits.

One of the primary attributes of Quartz Crystals is that these living mineral forms can receive, store, recover and transmit information. Your scientists discovered this two decades ago when they succeeded in putting a three dimensional image of the Mona Lisa into a quartz crystal and then retrieved it. Your main stream academics have long recognized the ability of crystals to transmit energy waves. Your early radios used crystal transmitters, and crystal or 'silicon' technology is the basis of your computers.

Masters, this is the only tip of the iceberg; or should we say the tip of the crystal!

The purpose of using a Phi crystal is to amplify and cohere the thought and energy that you wish to direct. Amplification is more readily understood, but coherence is a somewhat more difficult concept for you in physics or Metatronic terms. Coherence is a crystalline aspect of light energy waves, with each individual wave in intelligent ordered harmony with every other one. At present your laser light is the most prominent example of perfectly coherent light. All the photons emitted by a laser have the same frequency and are in harmonic crystalline phase in quantum states. We tell you that Phi Crystals are capable of forming and emitting a unique blend of intent-coded light in perfect coherent union. This may not be easy for you to grasp, but it is an important axiom, a key aspect of a 'new' paradigm as you enter the Ascension within the Quantum Crystalline Field. Coherency enhances manifestation of appropriate intent, you see?

Question to Metatron: Can you describe and explain the various benefits of Phi Crystals in the Vogel Cuts of 12, 33, 13-20-33, 88, 108 and 144 as well as the 'Dream Cuts?'

AA Metatron : *Indeed. First we will say that the Phi Crystals in the Vogel format are an Atlantean-Arcturian technology that utilizes specific base end angles to activate, vastly refine and greatly amplify the sentient characteristics of the quartz itself. There are many different kinds of benevolent frequencies available with Phi Crystals and of very, very different kinds of vibration, not what you would find in your naturally formed quartz. The Atlanteans and Arcturian Masters developed and understood the vast and complex Phi Technology of Crystals. Indeed, some of the Arcturian-*

Atlantean construct crystals placed by them in key nodes on the planet were intricately designed to emit Phi frequencies and are awakening now. However, those formed naturally in time in the Earth do not have that same refined vibration, you see?

So it is the Phi, the Golden angles and sacred lines that create the revelation of the pulsing Crystalline energy, activating and merging with the Quantum Crystalline Field to revere that which is holy, to reveal the highest symbol, the highest vibration of those which are the deity energies of crystalline light from the sun and the great central sun. So the element of Golden Phi, then, originally has come from the Sirian sun, and here it has founded itself upon the Earth. And where these omnipotent Phi Arcturian crystals are present, true deep and wise vibrations are found also.

But we note that not all quartz is of such vibration. Indeed some of the Master Atlantean Crystals are, and we have told you previously of these and that they are being awakened now from dormancy.

Natural quartz formations generally do not hold such a high frequency and are neither phi in resonance angles or of gold alloy. Some do but not most. Those crystals of the gold-silicate alloys were originally used in both the crystal skulls and the Master Atlantean crystals that carry the highest field. Sparse pockets of these in natural form do exist in Brazil, Tibet and Arkansas deep below the surface, but these are very few and have not been truly accessed.

Natural quartz crystals are hexagonal and release a spiraled phi energy that is emitted most powerfully through the point. Because few naturally formed specimens are perfectly symmetrical, the Arcturian and Atlantean Scientist Priests of the Atla-Ra developed a means of precise cutting of quartz in a double terminated format with equally symmetrical faceting to allow the Universal Life Force to be emitted through the crystal. In so doing, they realized that the Chi Energy from the human biology can be transferred into the crystal and better refined and spiraled to a focal precise position on the upper point.

The base angle was determined to most fluidly receive and refine chi energy with higher dimensional Phi energy when faceted to the angle of 51 degrees and 51 minutes, which is the angle of the Quad-Pyramidal Octahedron. It is the sacred angle that transmutes Celestial Energy into the Bio Mineral Sphere of the Primary Polar Earth Plane. It is the THOTHIAN METATRONIC angle used in your Great Pyramid. Some pyramid academics refer to this angle and measure it as 51.8 degrees, in fact it is 51 degrees, 51 minutes and 51 seconds. Do you see the Mastery in connecting the octahedron as such to as above so below. It connects both worlds, Dear Ones, and indeed is divine in its sacred resonance.

There are other precise angles that can be used for specific purposes both on the base and tip that have not as yet been 'rediscovered,' but this will be given in the appropriate timing. In the NOW, the 51.51:51 degree angle is most beneficial for the awakening of mankind to his initial phase of TRUE SELF realization. The geometrics draw a precise sacred energy field to it. It draws the human field to it as well, you see?

The one you refer to as Marcel Vogel, Khirron, recognized the synergy between crystals and water. The Phis create a synergy that belongs quite nicely to that which blends water and light, the two alchemical elements, you see? The blend of the two – one that both hardens and softens. It softens the soul. It strengthens the spirit, you see?

Now, all PHI Cut (Vogel) crystals have seven common primary properties that occur within them. Natural quartz crystals contain some of these properties, but these are greatly refined and amplified in the Phi cut units. These are the abilities to:

1. *Amplify Thought Frequency*
2. *Enable Focus for Manifestation*
3. *Receive and Store Thought or Intent*
4. *Transform or Refine Light Wave Frequencies*
5. *Transfer Thought Energetics or Thought Forms*
6. *Merge Frequencial Energies*
7. *Emit a Frequencial Field*

Now, per your query we will offer brief descriptions of the key benefits and traits of the requested faceting, but do note that there are numerous other ratio combinations and faceting models that are not listed or herewith inquired:

Double Terminated 12 Facet Phi Crystal (Vogel): *The 12 Facet Vogel works on the level of humanity and accesses, activates and harmonizes the 12 strands of DNA and the 12 electromagnetic patterns of consciousness within duality. The 12 Vogel Phi fortifies the human EMF and enables the user to amplify thought frequency. In truth this occurs with all PHI Cut Vogels, and the greater the number of facets the greater this quality is amplified, as faceting enables more light and thus more energy. The 12 initializes the awakening of man on the physical conscious level and creates the desire for more understanding. It brings in the Ruby Ray of Strength, Will and Perseverance.*

Double Terminated 33 Facet Phi Crystal (Vogel): *The 33 Facet Phi works on the level of humanity and assists mankind in clearing obstacles that impede advancement into the Christos. The 33 advances the desire for clarity, for greater compassion and for an understanding of the unity of all mankind. It enhances the spirit of brotherhood and brings the desire for unconditional love. The 33 taps into the 5th dimension. It brings in the Emerald Ray of Harmony and Healing. It resonates with the energy of the Elk and Deer.*

Double Terminated 20-33-13 Facet Phi Crystal (Vogel): *This special ratio crystal taps into that which is termed the Melchizedek Order of Alpha and Omega from the stance of humanity. It introduces the Divine Feminine (Emerald Ray) and the Platinum Ray. It carries all the characteristics of the Vogel 33, with special ratio characteristics. It is inspired thru the channel by the work of Enoch, of Kamadon. His energy is herewith imbued in this unique ratio. It may be termed the Melchizedek vibration. It is nurturing and healing and taps into the ratios and frequencies of the 4th, 5th and 6th dimensions. From a healing perspective, it is especially effective in working with and releasing emotional trauma and in allowing for the clearing of these blockages to establish the true Soul's Divinity within the SOVEREIGN SELF. It assists in clearing karmic relationship issues. It heals and enables the clear functionality of the glandular system. The Kamadon ratio is very effective in healing work and in activating the Goddess Energy of the sacred Platinum Ray.*

Double Terminated 44 Facet Phi Crystal (Vogel): *The 44 also operates on the level of humanity. It connects the individual to the concept of the oversoul. It makes known the unknown and compels greater investigation. It restores memory and offers the entry level of Earth completion. It is* *healing in the physical aspects of the joint inflammation and arthritic conditions. It is helpful in balancing purpose and energy, helpful in stimulating brain activity and circulation. The 44 Vogel Phi assists in communication with the Animal Kingdom and the sacred totem nurturer, the feminine spirit aspect such as the Buffalo as designated by the indigenous of North America.*

Double Terminated 77 Facet Phi Crystal (Vogel)*: The 77 Vogel is quite unique in its frequency. It vibrates extremely well with the Devic Kingdom, that of the Fae and of the Cetean, the dolphin. It opens the psychic channels into elemental kingdoms. It brings in the Pink Ray. It enhances the Spirit of the Eagle and winged aspect of the Animal Kingdom and Totem.*

Double Terminated 88 Facet Phi Crystal (Vogel): *The 88 crystal operates on the level of the oversoul and introduces multidimensionality. The 88 operates as a portal, capable of transmitting and receiving energies from the higher dimensions as well as enhanced thought manifestation and transferal. It is the Crystal of the Saint Germain energies. The 88 links the past to the future in forming the union of the NOW. It links and harmonizes the two hemispheres of the brain. It is the initial crystal of the Merkabic Transformation and introduces the first level of the Merkivah. The 88 correlates to the Blue Ray of Wisdom and Conviction. It correlates to the energy of the kundalini, dragon and transformational aspect of the sacred snake totem.*

Double Terminated 108 Facet Phi Crystal (Vogel): *The 108 represents the 9 of completions and connects to the Violet Ray and to the energy of Archangel Michael. It is also a portal, enhancing higher dimensional communication and* *multidimensionality. From a healing perspective, the 108 is especially capable of fine tuning the chi energy throughout the body and of working with the nervous system. It is especially effective in balancing energies, reducing over intensity, anxiety and panic disorders. The 108 amplifies life force in those with a lack of vitality and chronic fatigue. The 108 is also connected to the electromagnetic grid, the icosahedron, and is especially attuned to the vibration of GAIA – vortexes, portals, ley lines and certain powernodes or infinity points. The 108 resonates with the Mineral Kingdom and Living Earth, and it enhances communication with same. It is the vibrational tool of the Earth-Keepers and is effective in connecting to specific portals for alignment to same; it is effective in working with GAIA. It attunes to the 20 Point Star Level of the MerKiVah. It corresponds to the spirit of the Jaguar and Puma.*

Double Terminated 144 Facet Phi Crystal (Vogel): *The 144 is the most potent of the available Phi Crystals. It is the energy of Thoth and the Field of Metatron. It connects to the Elohim level of the 12 squared, a key frequency for mankind's ascension. The 144 also directly connects to the 144 Crystalline Grid and Quantum Crystalline Field. It is very beneficial in achieving the advanced Merkivah State to the 24-pointed star. It is the frequency of completion. It assists the union and harmony of all lifetimes created by the oversoul into the merged oneness of Mastery. From a healing level, the 144 is especially helpful in unlocking blockage and disease in the kidneys, liver and spleen. The 144 brings in all the rays and correlates to the Golden Ray of Omnipotence. The 144 is in itself a portal that enhances channeling and what may be termed celestial and dimensional travel. It is a 'contextual' portal, a time gate that can take one to the past, to the future or even to another day within your present. It is* *a doorway into parallels that can take you to another paradigm within your current reality so that you can experience it differently. If you are uncertain as to a given direction or decision and wish to experience it from a different perspective, a portal, or in this case a window, it can help you do just that. A portal can also be experienced as energy without expression, because experience does not require form, only context. A contextual portal has the advantage of allowing you to experience an event horizon whether it is the thought vision of it or actually manifesting it into being and living it. In the contextual portal both are equally valid, equally tangible from an experiential context. It is Merlin's Wand, the Philosopher's Stone of Alchemy.*

Now we will say that the Phi Vogels, in your terms, of the 88, 108 and 144 are the levels of Mastery, of the MerKiVah. These each access portals, with increasing frequency yet each within a unique spectrum. So we will clarify that one cannot truly employ the energetic frequency field of an 88 within the 108, nor the resonance of a 108 within the 144, or vice versa. It is not that way, you see. Each is unique through the faceting harmonic.=

Each opens a unique stance, stances that stack upon one another yet do not merge. Do you understand? Portals are utilized from different perspectives. There are many realities from which to choose, and there are many dimensions to explore. Densities within dimensions offer even more variety and infinite ways in which to explore and experience life. Portals are opportunities to experience yourself as beyond your everyday self. They activate expansion within and without and allow you to choose a reality rather than to have one chosen for you. Experiences beyond the 3rd dimension are best accomplished when desire, creativity, active intelligence and participation are all present. These conscious transformational experiences are available and accessible by you, but remember that they are possibilities, not requirements. They are not prerequisites that you must accomplish on your way here or there. They cannot gain you entry or reentry to that which has always resided within, but indeed they can assist you in experiencing these if and only if your energies are of a specific level of attunement.

Just because one goes to the energy field of a portal or time gate does not insure that one will frequencially experience either. Is this understood? The light quotient of the seeker must be of a high enough level to allow the merging and to afford the opportunity of such. As you humans are fond of saying, you must learn to walk before you can run. These are tools and potent ones in the hands of those seekers on the Path who are ready.

We will add that all modalities of true Phi Crystals, especially the Dream Phi and Vogel Phi (and there are indeed other cuts that have not emerged), are powerful instruments that operate in higher as well as infrared (below visible) frequencies of light and multidimensionality. Indeed many of you will instantly recognize them. Those of you who have used these Master Healing Crystals in past Atlantean sojourns will be immediately drawn to them and have an innate understanding of their capabilities.

These tools however require study, require interaction. They are not 'ready to use' per say. There must be the appropriate and deliberate effort to communicate, integrate and synergize with the living consciousness of each Phi unit. And this effort is in the vibration of high intent, of the desire for growth, within love and healing with compassion. Phi Crystals require a suitable time for activation, integration and charging. They must then be programmed with intent. One cannot achieve the desired results without taking the initial steps of melding beforehand. And a key part of the pre step is having sufficient knowledge as well as compassion (light quotient) to be able to raise a frequency high enough to enter their resonate field.

They cannot and should not be used for any purpose other than the highest good. They will not function in any intent or thought other than the highest good.

Question to Metatron: Can you speak on Dream Crystals, how are the different from Phi Vogels and what are their functions in the 3 Gate, 5 Gate and 7 Gate models?

AA Metatron: *The base angle of 51.51:51 degrees is the same on both. That which differentiates the 'Dream Phi,' to use your nomenclature, is the convex and concave cuts on the body of the Dream model, as well as the 'charging filter' on the base tip. The charging filter, often referred to as a flower, is a precise cut used primarily on Dream Phis but can equally be used on Vogel Phis. It works as a preamplifier.*

Now, keep in mind that each crystal fashioned from natural quartz is individually unique. The usage that we offer in both Vogel Phi and Dream Phi are somewhat generalized in primary application. Some forms of quartz, such as smokey and rutilated will have additional properties. In truth, each piece will have additional and specialized aspects. The user then is required to 'test drive' each to learn the characteristics of each. Water clear quartz is different from included quartz.

Using a Dream Phi with a Vogel Phi in tandem is highly synergetic. The Vogel Phi is in a sense the macro and the Dream Phi the micro. One the outer world, the other primarily the inner. The two dance with light in different rhythms.

The inner wave spiral and rotation of the beam in the Dream Phi is opposite to that to the Vogel Phi. The Dream Crystal by means of its convex concave structure has what may be termed as a clockwise spin and the Vogel Phi counterclockwise, although that description is somewhat metaphorical for use in explaining their opposite spins of light. It is not in truth a simple clockwise counterclockwise action, but the concept is a valid simplified image for this descriptive purpose. The Phi true action with light is far more complex.

We can say that the combination of the two succinct energy models combine two 'wholes' and form a greater wholeness. Combining then the inner and outer, the conscious and subconscious, the expressed and unexpressed, that felt and that seen. For healers working in the shamanic tradition, Phi Dream Crystals are an especially beneficial tool, a refinement for extraction work. Dream Phis emit a light field resonance that removes negative energies, attachments and obstacles from the auric field, emotional body and indeed the physical body.

All Dream Crystals take the cleansing of your white sage smudging to the level of light and complete the purification in that tonal context. These crystals once attuned are very effective in repelling untoward or 'negative' thought form energy.

The Dream Phi interacts in the interior worlds of thought, worlds devoid of space, worlds of dream and vision, of imagination and parallel. These dimensions exist and are fully valid in their aspect. Equally, there are aspects of you that would rejoin with other aspects of you and in the process become greater. If you add a drop of water to a glass, it will rejoin with itself and to the ocean again. So always when there are grander aspects of your beingness, you will rejoin with them, and you will change, expand in the process. It is appropriate, it is the expansive nature of your purpose and paradigm.

***Double Terminated Dream Phi 3 Gate:** The 3 Gate Dream Phi works on the level of humanity and the trinity of beingness. It amplifies harmony within the body, mind and soul through compassion. The 3 Gate Dream Phi Crystal has the ability to connect the individual with higher waves of information and energy. The energy from the 3 Gate radiates around the emotional, heart and crown chakras as it balances physical, intellectual and spiritual fields. It reconnects all the passageways of the matrix and allows the brain to operate more smoothly. In healing mode, it works in the extraction and repulsion of negative aspects including fear, doubt, greed, self centeredness and self righteousness. It assists in ego imbalance and physical imbalance in order to gain a sense of greater unity and harmony. It is the Phi for those entering new pathways of knowledge. It corresponds to the energy of the wolf.*

***Double Terminated Dream Phi 4 Gate:** The 4 Gate Dream Phi is based on Cabbalistic Wisdom and the Tree of Life. It is the progression of the 3 Gate. It is especially compatible with the 44 Vogel. The 4 Gate can be used for gaining understanding and removing false beliefs, breaking ties from group controls and controlling relationships. It enhances the desire to work well with others and reveals blockage to same.*

It works very well with the Animal Kingdom and is helpful in healing of the four legged. It is very helpful in communication with all Animal Totems.

Double Terminated Dream Phi 5 Gate: *The 5 Gate Dream Phi works especially well with the Melchizedek or Kamadon Vogel in bringing in a very powerful feminine energy. It correlates to the Emerald and Platinum Rays. This shape releases obstruction around deep trauma. It opens the way to the happiness and euphoria that is pure and impeccable in our transparent and sincere core within the etheric heart chakra. It is especially harmonious with the Divine Feminine with the aspects of mercy and nurturing Magdalena, Isis, Lady Faith and Quan Yin. It reveals the lack of Self Love, and assists in extraction of self denial, mentality of lack and the core issues around lack of energy and initiative. It assists in regaining momentum and purpose.*

Double Terminated Dream Phi 6 Gate: *The 6 Gate Dream Crystal is very effective in assisting in relationship issues. It reveals the inner motives of the user and assists in purging that which is in conflict with the higher self. It assists in revealing and removing that termed 'Spiritual Ego.' Likewise it effectively repeals energies of jealousy, control and anger directed at the user from another. It reveals likewise these conflicts within the self. The Six Sided Dream Crystal is often experienced as a total body energy radiating outward to expand the energy field of the user. It especially purges and balances the second, third, fourth and fifth chakras so that the mind and body are more connected and in better communication. It is the sword of Michael, and indeed it cuts both ways, exposing our own as well as others' true motives and conflicts. It must be used wisely and compassionately. It adds the information, feelings, and "software" necessary for* *operating the matrix that has been reconfigured and healed. Much of the experiential information from Atlantis radiates from these Six Sided Dream Crystals. It is expected that the overall effect of this Dream Crystal will be to enhance and expand the doorways of knowledge to Atlantean technologies and spiritual practices.*

Double Terminated Dream Phi 7 Gate: *The 7 Gate Dream Phi specialty is bringing perfect order, completion, reconciliation, unity and connection with eternity. It brings all other healing energies together and connects you with spirit. The 7 Gate is the Platinum Ray and correlates very well to the 88, 108 and 144 Vogel Phi. The 7 Gate Dream Crystal receives and merges wisdom from all spiritual sources and views, from all peoples, ethnicities and cultures. It is the crucible that shows what is pure and what is not in spiritual path. It repels untruth, it repels control and enhances freedom and power of choice and will. It reveals hidden prejudices and self concealed vices within the user for extraction of same. It enables the MerKiVah. It enables better contact with the conscious grids, the Reshel and planetary grids.*

Double Terminated Dream Phi 12 Gate: *The 12 Gate Dream Phi is the highest frequency of the Phi Dream. It corresponds to the Purple, Golden and Platinum Rays of the Ascension. It correlates to the Metatronic Cube and 24 Point Star of the Merkivah. It works especially well with the 144 Vogel. It offers protection in the higher realms of inner micro dimensional travel and assists in channeling. It stimulates the intellect and blends the hemispheres of the brain. It connects to the 144 Crystalline Grid, the Golden Sun Discs, and it activates the 12 strand DNA. It opens Celestial Portals and connects to the Elohim.*

Question to Metatron: Minerals, Gems and Phi Crystals are often expensive, and seemingly not available to all. Why is that, and are these necessary 'tools?'

AA Metatron: *First of all, gems, minerals and noble metals are in truth quite available to anyone that seeks them, to anyone that would draw them to self. You must recognize first that the Mineral Kingdom is alive and sentient in all its myriad varieties of expression. Within the Mineral Kingdom there is indeed the spark of Divinity, of Divine Consciousness that is quite equally aware in certain aspects, as is humankind. This is a fact that is as yet unrecognized by most of humanity, yet it is quite true.*

It is only the mentality of lack and the non recognition of the crystalline and mineral kingdom as being sentient that keeps anyone from manifesting these. Your societies and cultural systems spend unquestioned amounts of energy or money in the acquisition of what may be termed luxury items and excessive amounts on dress clothing, dwellings and vehicles. Yet they do not recognize that energy exchanged for certain forms of the mineral kingdom offer them great advantages in health, consciousness and manifestation amplification.

Certain of your civilizations recognized the Mineral Kingdom as being sentient, and as a result of that key acknowledgement they were able to communicate and interact with the mineral and crystalline beings. Accordingly the Mineral Kingdom offered itself and made itself readily and bountifully available to mankind. And, Dear Ones, the Mineral Kingdom responded to their desire, to their acknowledgement, to their call and made itself available to that call, to that seeking. Indeed it is an example of the Law of Attraction.

The same law is valid today, in your present. Even though you do not generally go to the fields and valleys to find your minerals, you are still very capable of sending out the acknowledgment to the mineral kingdom and manifest to attract to you what you need, what you desire to assist you. The mineral itself, the sentience of the mineral once acknowledged will indeed make itself available to you. That is quite true, believe it!

The paradox is that if you do not believe it, indeed that is the reality you will manifest. If you do not recognize the mineral kingdom is alive and sentient, then you will not be acknowledging its sentience. And acknowledgement, belief is essential, you see.

Unfortunately, the mass of humanity views gems and precious metals as simply decorative jewelry, vanities and investments. Therefore they are viewed by many as expensive embellishments. Those who are aware of their true nature, purpose and energy benefits then understand that they can and do enhance physical health and that they calibrate and generate higher frequency energy fields that can offer myriad advantages to those that understand them as such.

So as to your question, are gems, metals and crystalline forms necessary. We will answer in this way. Your cars, trains and planes are not truly necessary, you can walk, and if you are an extraordinary Master you can astrally bi locate. But in the interim, the tools you have manifested make transport a lot easier.

Some humans on the spiritual path remain limited by their mentality of lack. This is manifested in both a fear of poverty and in the belief that abundance is an aspect and indulgence of materialism. Both are false beliefs that do not serve you. Abundance offers itself and is available to all; the paradox is that you must believe that abundance is not only available but is equally a path of spirit. Once the laws of manifestation arte realized, Masters, you can create and acquire what you desire. Only be sure it serves your greatest good.

So indeed these tools, these living crystalline forms are available to those that are aware. And the truth, whether you accept it or not, is that the crystalline forms are energy generators, frequency portals and resonant shields that offer myriad benefits.

Question to Metatron : Please explain the proper manner of caring for, charging, and programming a Phi Crystal?

AA Metatron: *The methodology taught by Marcel Vogel is available to all and is the accurate modality. So we will accordingly not delve too deeply into information that is accurate and already easily available to you.*

We have described in the (above) discourse the requisite belief and importance of recognizing that all crystalline forms have sentient consciousness. Initial synergy and communication with the crystal is achieved primarily through transferal of the frequency of love into the crystalline form and asking for its assistance. This should be done daily with meditation and the use of deep breathing exercises to raise the vibration of the user. When the vibrational frequency of the theta wave is achieved, imbue the crystalline form through visualization. Imagine a beam of love from the heart chakra directly into the crystalline being. When a return flow of 'love' is felt, the synergy is achieved. In many cases there will be an electrical sensation felt in the arms of the user that flows upward into the heart. The actual sensation may vary with the user. For some it may be simply a knowing, for others a physical tingling, but the return flow must be unmistakably sensed by the user.

Once synergy is established, the infinity breath exercise should be employed in meditation whilst holding the crystal in order to establish intent or programming. Once the deep breathing takes one into the theta state, the crystal will signal when the field is appropriate for receival of intent.

At this point the user should lightly stroke the body of the Phi crystal with the fingertips to charge the Phi while sending the intent of the particular usage. Transfer the visualization of what is requested. Whether this is to assist in multidimensional awareness, self purification, protection or healing for self and others, the specific intent should be clear and imbued into the Phi.

The process of healing, the extraction of obstacles or disease is done only with the permission of the one receiving the healing. The Phi is then held with the female base in the palm of the primary hand of the user and circulated over the area of focus. For Vogel Phi Crystals, the circulation should be counterclockwise. For Dream Phis, clockwise. Visualize and speak the desired result. Continue this process until a clear response is received that the task has achieved maximum energy for the session. To seal the opening upon completion, circulate the opposite direction until the field is closed. Usually three to five circulations will reseal the field.

The most effective method of cleansing a Phi is through vigorous brief shaking in the hand while visualizing intent of removing unwanted energies and through the use of white sage smudge. Running water can also complete this process, but care should be taken to insure the temperature of the water is within ten degrees of the ambient temperature of the crystal. Rapid contraction or expansion due to great temperature variation can create inclusion cracking in the quartz structure.

Closing: The Changing Paradigm and Sacred Science

Masters, there are differing sciences, different physics for each dimension, just as there are different approaches and paths of science and physics available to mankind in his current reality. Man has omitted the divine from science in his mainstream thinking. Science is indeed a sacred aspect of the whole.

Mankind has largely ignored sciences that would have led to very different concepts thus far. There are alternative approaches to physics that reveal much more about transport, transmutation, bi location and locomotion than your accepted mainstream science understands or wishes to understand. Had the human species gone into certain mental disciplines as thoroughly as it has explored exterior technological 'laws,' your knowledge, means and resulting understanding of dimensional and crystalline reality aspects would be vastly different and far more efficient in scope than it is now. You have embraced the external and to some degree dismissed the capacities of your 'internal abilities' within divine consciousness. Crystalline knowledge is a key part of the emerging 'new' paradigm, although in truth it is an ancient knowledge, forgotten but not new! Now, when mankind decides to devote mainstream study into what is termed the 'mental' science of dimensional and crystalline aspects of transport and bi location, and indeed it is a science with laws that can be learned, practiced and fine tuned, then visitations into parallels and vectors within time and space and inner worlds will become less accidental and occur by design, by plan.

Once mankind learns and masters 'mental physics' then he will be liberated, vastly liberated from the filtering illusion, the duality camouflage of physical pattern. Indeed you are just starting to understand how the Merkabah and Merkivah, when tuned by mind as the builder into the crystalline field unlocks this key. This will be greatly enhanced as your 144 Grid and its networking of crystalline energy completes its unfolding. The grid will offer great opportunity of advancement to those of you willing to devote focus to this divine science of 'mental' manifestation and what you term astral travel. You see, all science must include the divine, and it is grossly omitted by your mainstream academics in the present.

The knowledge and utilization of Phi Crystals and indeed properties of all crystalline forms is an intricate and essential part of this aperture into the vast Void, into inner and outer worlds of the celestial and indeed Divine aspect of the Crystalline Ascension.

Phi Crystals are precision instruments, they are not for everyone. They are not simply crystals nor do they serve the mere curiosity of those who obtain them as vanities or novelties! They call to those who are ready. They offer themselves to those who are responsibly prepared to devote serious focal study and attention to their profound use and divine mysteries. These are sentient forms of divinely intelligent consciousness whose sacred geometric angles unlock dimensional gates that allow the prepared seeker to dance with creative light force and source. Phi Crystals offer extreme advancement to those with sufficient Light Quotient to activate in synergy with the Matrix of the Cosmos.

I am Metatron, and I offer you these truths in love and in wisdom. You are Beloved.

... and so it is. AA Metatron via James Tyberonn

Appendix ~ Activating the Vogel Phi for Healing

Now, in this sojourn it is well for the entity to redevelop the ability to use the Phi Crystals, the Vogels for healing of the human aura, the human electromagnetic field, both for application to self and to others. The objective then is the creation of the 'harmonic field.' The key instrument is the Phi Crystal. Use the Phi with toning, with tuning forks, with hemi-Sync audio patterns. And so we share this application for creating the affinity, healing field, understanding and application of the Phi Crystal.

The following serves as a basis, and we note that you not only can but should rely on your intuitive knowledge that will indeed emerge in time in order to alter the methodology to your unique application. The speed of unraveling will depend then on the amount of effort you apply in this study and on the trusting of self:

1. *Spend time with the crystal. Speak to it, for indeed it is alive and fully divinely sentient. Acknowledge it as a sacred ally, as a trusted soul friend. Ask the crystal to work with you in healing, in the highest good.*
2. *Initially stroke the crystal with the fingers, and tone the Ohm sound. Open every chakra, beginning with the crown down to the root and back up to the heart. Send love directly from the heart chakra and visualize it filling the crystal with emerald green light. Ask the crystal to acknowledge you and sense the response. It will come as an energy pulse from the crystal that you will recognize as an energy dynamic. It may feel as a sensation running up the arm into the spine and heart or as simply a mental understanding, a responsive flash to you from the sentience of the crystal. Spend at*

least three weeks doing this for a period of not less than 20 minutes per day, or until the energy reaction is sent back to you and recognized. Visualize and build a field of green energy that encompasses your energy field (MerKaBah – Star Tetrahedron). Continue until the MerKaBah is pulsing in vibrant vivid green.

3. *After completion of this field in the three week building of the healing field, hold the Phi Vogel in the primary hand (in the right hand if you are right handed, in the left hand if left handed). Place the female or wide base of the lower receiving end of the Phi crystal in the center of the open palm and close the hand around it. Stroke the body of the crystal with the thumb. Tone the Ohm, send in the heart energy through the palm. Wait for the acknowledgement response from the crystal. Then visualize the green 'heart love' energy spiraling through the crystal until the crystal warms noticeably in the hand.*
4. *Energy can be sent from the crystal in various methodologies. To heal or energize an auric field, first ask and receive permission from the one receiving the energy, even if it is self for self healing. To open the field, circulate clockwise until a trigger pulse response is felt from the crystal. At that point change the pattern of movement to the infinity, figure 8 pattern in clockwise motion. Then visualize the green healing energy coming from the God Source or Higher Self, through the crown to the root and back to the heart, from the heart to the palm and into the crystal. Then visualize the green healing energy spiraling out of the crystal's outer fine point, exuding, projecting in spirals from the male end. Then tone a unique sound from your divinity that you intuitively channel and know to be complimentary to the healing. Continue until the*

Phi crystal triggers you that the session is complete. Once the healing or energizing is complete, and the crystal will signal you, then reverse the infinity pattern of the crystal to counterclockwise. Wait for a trigger, and then circulate in 360 degree circles counterclockwise to seal the aura. Stop when the crystal pulses, usually after 5-20 circles. In time you will be able to optionally vary from the initial suggested modality to that which is distinctly your own. The tones you create will vary during healing according to what is needed. Trust this.

5. *After each session, acknowledge and thank the crystal. Shake it firmly, then sage smudge it and run it through ambient temperature water. This clearing should also be done before the session begins.*

The cleansing, healing and energizing process for the Dream Phi Crystal is the same, except that the circles are rotated in the opposite pattern (clockwise to open, counterclockwise to close). As the Dream Phi works in inner dimension, the Phi Vogel works in outer dimension.

The process for time travel, dimensional travel and channeling is more appropriate for the 144 Vogel Phi (not the Dream Phi). The 88, 108 and 144 will suffice, but the optimal is the 144 Phi Vogel. The process for this is the same as in step 1, it varies slightly in step 2 and has a unique process from steps 3 to 5:

Channeling Meditation, Time Travel with 144-Vogel Phi

1. *As described above.*
2. *Initially stroke the crystal with the fingers, and tone the Ohm sound. Open every chakra beginning with the*

crown down to the root and back up to the heart. Send love directly from the heart chakra and visualize it filling the crystal with emerald green light. Ask the crystal to acknowledge you, and sense the response. It will come as an energy pulse from the crystal that you will recognize as an energy dynamic. It may feel as a sensation running up the arm into the spine and heart or as simply a mental understanding, a responsive flash to you from the sentience of the crystal. Spend at least three weeks doing this for a period of not less than 20 minutes per day or until the energy reaction is sent back to you and recognized. Visualize and build a field of green energy that encompasses your energy field (MerKaBah – Star Tetrahedron). Continue until the MerKaBah is pulsing in vibrant vivid emerald green.

3. *After completion of this field in the three week building of the healing field, hold the Phi Vogel in the primary hand (in the right hand if you are right handed, in the left hand if left handed). Place the female or wide base of the lower receiving end of the Phi crystal in the center of the open palm, and close the hand around it. Stroke the body of the crystal with the thumb. Focus on the crown chakra. Visualize it in brilliant purple violet. Tone the 'AAEE' sound, actually verbalize it not just hear it mentally, actually chant the tone (Ayyee), and then send in the crown energy through the palm into the 144 Phi Vogel. Wait for the acknowledgement response from the crystal. Then visualize the green 'heart love' energy spiraling through the crystal until the crystal warms noticeably in the hand.*
4. *Direct the Vogel Phi Crystal to the Self , centered in the MerKaBic Field. Place the MerKaBah centered around you, and spin it counterclockwise. Then go to the 2nd phase of the MerKaBah and rotate the lower tetrahedron clockwise as the upper one continues in counterclockwise*

motion. Next, transform the Star Tetrahedron to the 12 Pointed Star of the initial phase of the MerKiVah. This is the stellated dodecahedron. Then spin it counterclockwise. Visualize it as rich purple. Tone the AAEE and do the Infinity Breath (figure 8 pattern).

5. *Transform the MerKiVah to the golden Ray color, rich brilliant Gold. Visualize it spinning faster until you begin floating into higher dimension.*
6. *See the MerKiVah enter into a lucid crystalline octahedron. Transform the MerKiVah into clear crystal, water clear. See yourself inside the MerKiVah as crystalline, transparent and brilliantly diamondesque clear. You are into the Crystalline body of the Unified Crystalline Field.*
7. *If your intent is to channel, call for your Higher Self* and *the Angelic Realm of Ascended Masters. Call them forward and receive.*

If your intent is dimensional travel, follow the above steps 1-6. Then:

7. *Extend a golden ray from each of the 12 points of the MerKiVah.*
8. *Apply the 'Golden 12 Squared Refraction,' where each one of the individual rays refracts and multiplies into 12 rays, so that 144 Golden Rays are now being emitted from the MerKiVah.*
9. *Express intent on where to travel, silence the mind from inner narration, and allow only imagery. Spin faster and then travel into one of the Rays.*

We will add that the process requires work, effort and discipline. It can take days, weeks or months to achieve depending on the level of commitment. Trust in Self.

About the Author

James Tyberonn worked as an engineer and geologist for 34 years in the energy sector. He is a native of Arkansas but lived and worked abroad for 31 years. He has traveled to over 55 countries in his geology work and lived as an expatriate in Brazil, India, Venezuela, Gabon, Congo, United Arab Emirates and Scotland. He studied gemology while in Brazil and speaks Spanish, Portuguese and French. He has had a great interest in metaphysics, sacred sites, spirituality and music.

He has visited over 200 sacred sites across the globe over the past three decades. He has been a guest speaker on numerous radio programs. He has mixed Native American heritage and has completed six 5-day vision quests and numerous shamanic journeys in Mexico and in South America. He has authored two books and is writing his third on the metaphysical healing properties of gemstones. In addition to writing, he is a recorded musician and dedicates time to charity fundraisers for the Cystic Fibrosis Foundation. Tyberonn began channeling Archangel Metatron in 2007 and is regularly featured in *Sedona Journal* magazine. He conducts seminars and sacred site travel throughout the Americas.